Big Data for Remote Sensing: Visualization, Analysis and Interpretation

Nilanjan Dey · Chintan Bhatt
Amira S. Ashour
Editors

Big Data for Remote Sensing: Visualization, Analysis and Interpretation

Digital Earth and Smart Earth

Springer

Editors
Nilanjan Dey
Department of Information Technology
Techno India College of Technology
Kolkata, West Bengal
India

Chintan Bhatt
Charotar University of Science and Technology
Changa, Gujarat
India

Amira S. Ashour
Department of Electronics and Electrical Communications Engineering, Faculty of Engineering
Tanta University
Tanta
Egypt

ISBN 978-3-030-07895-9 ISBN 978-3-319-89923-7 (eBook)
https://doi.org/10.1007/978-3-319-89923-7

Softcover re-print of the Hardcover 1st edition 2019

Printed on acid-free paper

This Springer imprint is published by the registered company Springer International Publishing AG part of Springer Nature
The registered company address is: Gewerbestrasse 11, 6330 Cham, Switzerland

Foreword

In this digital era, the size of the data involved in countless applications of our society has been increased substantially. Therefore, new computational methods, algorithms, and infrastructures are highly demanded in order to handle such sets of big data more efficiently, mainly in numerous real-time applications, requiring less powerful computational resources, so can be processed by common computational solutions.

Among the various areas where big data sets have become common, the ones related to Remote Sensing and information and communication technology are foremost, since the datasets involved have reached huge dimensions, which makes exceptionally complex their visualization, analysis, and interpretation. Therefore, this book assumes imperative and timely significance for these areas by presenting, discussing, and suggesting applications, infrastructures, methods, and techniques to overcome the present drawbacks. Five chapters are included in this book addressing issues of big data in e-health, aerial and satellite imagery, many-particle systems, earth science, and Remote Sensing.

The three editors of this book are well-known academics and researchers: Dr. Nilanjan Dey is Professor at Techno India College of Technology, is editor of several international journals, editor of various books published by renowned publishers, is co-author of numerous articles published in the most respectable journals and conferences, has supervised several M.Sc. and Ph.D. theses and organized many scientific events. Dr. Dey's research interests include medical imaging, soft computing, data mining, machine learning, rough set, mathematical modeling and computer simulation, biomedical systems, robotics, information hiding and security. Dr. Chintan Bhatt is Assistant Professor at Charotar University of Science and Technology, is co-author of several articles published in journals and conferences, has organized special sessions devoted to Internet of Things and big data, and his research topics are: Data Mining, Networking, Big Data, Internet of Things and Mobile Computing. Dr. Amira S. Ashour is Professor and Vice Chair of Computer Engineering Department, Computers and Information Technology College, Taif University, is editor of journals and of books published by international publishers, and co-author of numerous articles published in the most well-known journals and conferences. Dr. Ashour's research interests include image processing and analysis, pattern recognition, signal, image, and video processing, medical imaging, biomedical systems, soft computing, artificial intelligence, and wireless and mobile communication.

At this time, I would like to congratulate the editors of this book for their amazing effort to tackle so timely and interesting topic and relevant knowledge, experiences, and guidelines that are going to contribute to overcome the current difficulties concerning the visualization, analysis, and interpretation of Remote Sensing big data. In addition, my special acknowledgment goes to the authors of the included chapters for sharing their expertise with the scientific community.

Finally, I am pleased to recommend this book to all students, researchers, and end users related to problems comprising Remote Sensing big data, since it will facilitate their endeavors in developing more competent solutions for such problems.

Porto, Portugal João Manuel R. S. Tavares
Universidade do Porto

Preface

The current advances in remote detecting (RS) and ICT technologies give birth to huge amount of remote detecting (RS) information. The RS information collected by a solitary satellite is drastically growing by a few terabytes for every day. As per the insights of Open Geospatial Consortium (OGC), the worldwide observation information would most likely surpass one exabyte. Particularly, the coming of the high-resolution earth observation period (EOS-4) has likewise prompted the high dimensionality of the RS image information. Remote detecting information is perceived as "big data" in some specific sense. In the meantime, the precise and cutting-edge data provided worldwide earth observations are changing the way that earth is interpreted. Clearly, expansive remote sensing applications overpowered with huge remote sensing information are viewed as data-intensive issues.

This book involves five chapters. The work *Big Data Approach for Managing the Information from Genomics, Proteomics, and Wireless Sensing in e-Health* by J. Demongeot, M. Jelassi, C. Taramasco aims to show that big data techniques can serve for dealing with the information coming from medical signal devices such as bio-arrays, electrophysiologic recorders, mass spectrometers, and wireless sensors in e-health applications, in which data fusion is needed for the personalization of Internet services allowing chronic patients, such as patients suffering cardiorespiratory diseases, to be monitored and educated in order to maintain a comfortable lifestyle at home or at his place of life. The information provided by the pathologic profiles detected and clustered thanks to big data algorithms, is exploited to calibrate the surveillance at home, personalize alarms, and give adapted preventive and therapeutic education.

The work *Aerial and Satellite Imagery and Big Data: Blending Old Technologies with New Trends* by J. Salazar Loor, P. Fdez-Arroyabe does brief review of RS history presented in section one. Then, basic properties, which are also challenges, of RS big data are concisely discussed. Volume, variety, and velocity are mainly described as characteristics of RS big data while variety, value, and visualization are primarily denoted as new challenges. The third section is concentrated on justifying the relevance of RS big data in today's society and the needs to integrate it with other kind of data sources to develop useful services. In this sense, a special section is dedicated to Copernicus initiative, and some case studies of specific applications are also shown. Finally, some general conclusions are presented paying attention to the spatial nature of RS big data, which gives it a special added value in the new digital era.

The work *Structure and Dynamics of Many-Particle Systems: Big Data Sets and Data Analysis* by Wolfram Schommers underlined that the structure and dynamics of many-particle systems determine essentially the properties of the systems in solid-state physics, materials science, and nanotechnology. The fundamental description of such systems is based on atoms or molecules that interact with each other. The physical background is discussed in detail. In particular, it is argued that the characteristic features and properties of such systems are already reflected by relatively small parts consisting of 10^2–10^7 particles. It is outlined that for all the systems used in solid-state physics, etc., no reliable analytical models exist and we have to recourse to numerical methods. The procedure is to solve Newton's equations of motion numerically using the interaction potential as input, and these fundamental equations of motion are expressed by coupled differential equations (molecular dynamics). The essential features of the molecular dynamics method have been discussed.

The focus of the work *Earth Science [Big] Data Analytics* by Mani Madhukar, Pooja is to analyze what exactly does big data mean in earth science applications and how can big data provide added value in this context. Furthermore, this chapter demonstrates various big data tools which can be mapped with various techniques to be used for experimenting earth science datasets, processed, and exploited for different earth science applications. In order to illustrate the aforementioned aspects, instances are presented in order to demonstrate the use of big data in remote sensing. Firstly, this chapter presents earth science studies, application areas/ research fields, and a brief insight on earth science data. Then various techniques implemented in this domain are elaborated, viz. classification, clustering, regression, deep learning, pattern recognition, machine learning, earth data analysis, and processing. Finally, it is shown how different tools are mapped to earth science datasets using ArcGIS to illustrate with experimental instances the inferences and patterns generated.

The work *Retrieval of Urban Surface Temperature using Remote Sensing Satellite Imagery* by Jinxin Yang, Man Sing Wong, Hung Chak Ho presents the Urban Surface Temperature (UST) retrieval with consideration to the urban geometric characteristics in different seasons, analyzing the effective emissivity and urban surface temperature. Emissivity is crucial for surface temperature retrieval. However, the cavity effects and thermal heterogeneity caused by complex buildings affects the effective emissivity over urban areas. In this study, the effective emissivity from ASTER products in different seasons was collected to study the thermal heterogeneity effects on the applications of temperature and emissivity separation (TES) algorithm on the UST retrieval. Thermal images of Landsat 5 in different seasons were collected for analyses, in which the retrieved USTs, with and without considerations to geometric effects, were compared and analyzed. Finally, SUHI estimates based on two sets of USTs and its impacts on SUHI intensity estimation at different seasons were also analyzed.

We hope this book introduces capable concepts and outstanding research results to support further development of big RS data for analysis, interpretation, and visualization.

Kolkata, India Nilanjan Dey
Changa, India Chintan Bhatt
Tanta, Egypt Amira S. Ashour

Contents

About the Editors

Nilanjan Dey received his Ph.D. from Jadavpur University, India, in 2015. He is an Assistant Professor in the Department of Information Technology, Techno India College of Technology, Kolkata, West Bengal, India. He was a visiting scientist of Global Biomedical Technologies Inc., CA, USA (2012–2015). Research Scientist of Laboratory of Applied Mathematical Modeling in Human Physiology, Territorial Organization of Scientific and Engineering Unions, Bulgaria, and Associate Researcher of Laboratoire RIADI, University of Manouba, Tunisia. His research topics are medical imaging, data mining, machine learning, computer-aided diagnosis, atherosclerosis, etc. He is the Editor-in-Chief of *International Journal of Ambient Computing and Intelligence* (IGI Global), USA, *International Journal of Rough Sets and Data Analysis* (IGI Global), USA, the *International Journal of Synthetic Emotions* (IGI Global), USA (Co-EinC), and *International Journal of Natural Computing Research* (IGI Global), USA; Series Editor (Co.) of *Advances in Ubiquitous Sensing Applications for Healthcare* (AUSAH), Elsevier, *Advances in Geospatial Technologies* (AGT) Book Series, (IGI Global), USA, Executive Editor of *International Journal of Image Mining* (IJIM), Inderscience, Associated Editor of *IEEE Access* and *International Journal of Information Technology*, Springer. He has published 30 books and more than 250 research articles in peer-reviewed journals and international conferences. He is the organizing committee member of several international conferences including ITITS, W4C, ICMIR, FICTA, ICICT, etc.

Chintan Bhatt is an Assistant Professor of U and P U Patel Department of Computer Engineering at Charotar University of Science and Technology. His Ph. D.* is in Computer Science. He has done M.Tech. in Computer Engineering, Dharmsinh Desai University, 2009–2011, and B.E. in Computer Engineering, Gujarat University (Charotar Institute of Technology), 2005–2009. His research interests include data mining, software engineering, networking, big data, Internet of Things (IoT), mobile computing. He received an award on 2015, Paper Publication at International Conference Award and Faculty with Maximum Publication in CSIC Award during CSI Annual Convention, New Delhi. He had organized special sessions on Internet of Things (IoT), CSNT, Gwalior, during

April 4–6, 2015, and on Internet of Things (IoT) and big data in IoT, ICTCS, Udaipur, during November 14–16, 2014. He is an active reviewer in the *Wireless Communications*, IEEE, and *Internet of Things Journal*, IEEE, as well as *Mobile Networks and Applications*, Springer. He has several publications in repute journals. He has edited four books in IGI Global and Springer, and he is currently editing two books under Elsevier.

Amira S. Ashour is currently an Assistant Professor and Head of Department, EEC, Faculty of Engineering, Tanta University, Egypt. She has been the Vice Chair of Computer Engineering Department, Computers and Information Technology College, Taif University, KSA, for one year from 2015. She has been the Vice Chair of CS Department, CIT College, Taif University, KSA, for five years. Her research interests are smart antenna, direction of arrival estimation, targets tracking, image processing, medical imaging, machine learning, signal/image/video processing, image analysis, computer vision, and optimization. She has published 10 books and about 70 published journal papers. She is an Editor-in-Chief of *International Journal of Synthetic Emotions* (IJSE), IGI Global, USA. She is an Associate Editor of *International Journal of Rough Sets and Data Analysis* (IJRSDA), IGI Global, USA, as well as of IJACI, IGI Global, USA. She is an Editorial Board Member of *International Journal of Image Mining* (IJIM), Inderscience.

Chapter 1
Big Data Approach for Managing the Information from Genomics, Proteomics, and Wireless Sensing in E-health

J. Demongeot, M. Jelassi and C. Taramasco

Abstract This chapter aims to show that big data techniques can serve for dealing with the information coming from medical signal devices such as bio-arrays, electro-physiologic recorders, mass spectrometers and wireless sensors in e-health applications, in which data fusion is needed for the personalization of Internet services allowing chronic patients, such as patients suffering cardio-respiratory diseases, to be monitored and educated in order to maintain a comfortable lifestyle at home or at their place of life. Therefore, after describing the main tools available in the big data approach for analyzing and interpreting data, several examples of medical signal devices are presented, such as physiologic recorders and actimetric sensors used to monitor a person at home. The information provided by the pathologic profiles detected and clustered thanks to big data algorithms, is exploited to calibrate the surveillance at home, personalize alarms and give adapted preventive and therapeutic education.

Keywords Big data · Genomics · Proteomics · Wireless sensing
E-health · Data fusion · Alarm triggering

J. Demongeot (✉)
AGEIS, EA 7407, Faculty of Medicine, University Grenoble Alpes,
38700 La Tronche, France
e-mail: Jacques.Demongeot@agim.eu

M. Jelassi
RIADI, National Engineering School of Computer Sciences,
Manouba University, 2010 Manouba, Tunisia
e-mail: jlassi.mariem.esti@gmail.com

C. Taramasco
Escuela de Ingeniería Civil en Informática, Universidad de Valparaíso,
General Cruz 222, Valparaíso, Chile
e-mail: Carla.Taramasco@uv.cl

N. Dey et al. (eds.), *Big Data for Remote Sensing: Visualization, Analysis and Interpretation*, https://doi.org/10.1007/978-3-319-89923-7_1

1.1 Introduction

The recent big data techniques are useful for compressing, clustering and modeling health data during the visualization, analysis and interpretation processes, especially in medical practice. The flow of data coming from e-health wireless sensing, namely from the systems of surveillance of elderly or chronic patients at home is dramatically increasing and needs to be treated thanks to the "new" big data tools (often a new combination of former classical approaches) (Fig. 1.1).

The present overflow of health data has as consequence to displace the barycenter of the information from hospital centered systems (such as the first developed in Europe, Diogène® in Geneva, Bazis® in Leyden and Crystal Net® in Grenoble) to the patient. This tendency is also observed in USA, where information systems like the e-blue button by Humetrix® are patient-centered and at least so used by paramedics and by the patients themselves at home, than by physicians. The information content of the medical file of a given patient was in the seventies about 500 ko of clinical and biological data. With the development of computerized medical imaging devices and their storage facilities (such as PACS, Picture Archiving and Communication Systems) in the eighties, this information content passed to 5 Mo, but, with the introduction of modern tools of genomic and proteomic studies (namely bio-arrays and mass or NMR spectroscopy), this content jumped to 5 Go and more. The genome indeed is coding for about 20,000 structural and metabolic proteins, plus about 230,000 immuno-proteins. By considering the regulatory networks in which genes and proteins are involved, even depending in a binary way on their concentration (reduced to the value 1 over a certain threshold of expression and 0 under this threshold), it is possible to manipulate trajectories of

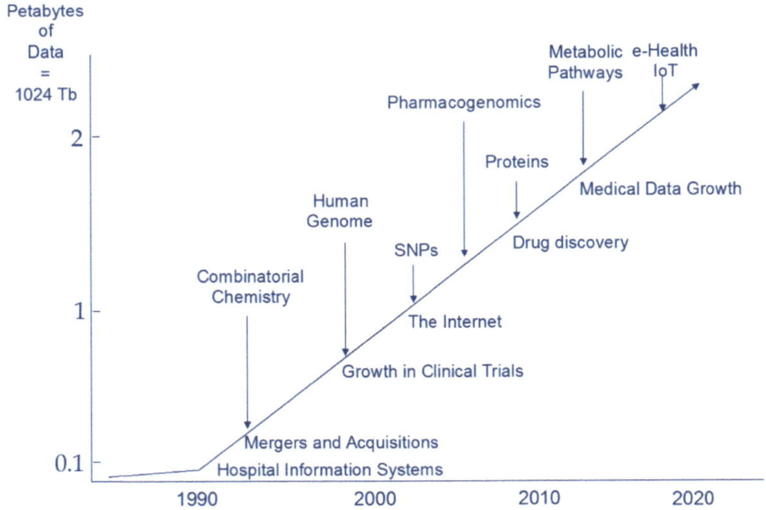

Fig. 1.1 Medical data produced pro year in a hospital receiving a total amount of 200,000 patients/year

data along the duration of a disease in a huge state space, the Boolean hypercube $\{0, 1\}^{250,000}$, whose size is $10^{75,000}$. These trajectories are depending on a set of dynamic rules relating the state of a patient at time t to his state at time $t + 1$. These dynamic rules are closely depending on the existence of relationships of various types (inhibitory or activatory) between the genes and proteins of the regulatory metabolic networks directly linked to the healthy or pathologic state of the patient.

Section 1.2 consists of a rapid survey of the big data approach in health, from the start of the usage of numerical data in medicine to the invention of the main big data operators, such as the Neural Network (NN) convolution and renormalization, used now for treating new e-health data. The history of these tools is simple: after a multiplication of data analysis tools from the seventies until the nineties (such as Spearman's factorial analysis, Benzécri's correspondence analysis, MacQueen's k-means, Diday's "nuées dynamiques", Hopfield's neural networks, Kohonen's maps, Hérault/Jutten/Comon's Independent Component Analysis, etc.), the successive introduction of powerful storage tools (from hierarchical to relational data base management systems, and from object-oriented to no-SQL data warehouses) pushed the medical data provider and user to invent new analysis tools for managing and processing the stored big data, these tools being often just a smart combination of previous mathematical algorithms of descriptive data analysis (mixed if necessary to classical tools of inferential statistics). After this survey, examples of recently invented big data algorithms are described, presenting successively a big data compression tool generalizing the wavelet transform (called "dynalet" transform), then a weak version of the order statistics (called "monotony signature"), after a clustering method, the Markovian final classes method (known as MCL), useful for clustering cardio-respiratory data coming from the surveillance at home of patients suffering cardio-respiratory diseases, and eventually, a method of modeling using an improvement of the classical kernel estimation method of the functional non-parametric approach.

Section 1.3 describes some redundant multi-sensing techniques necessary for example to anticipate changes in the ordinary life of a person at home in terms of (i) actimetry for detecting for example the appearance of a progressive motor impairment, (ii) sensory control for measuring the loss of visual or auditory abilities and (iii) vegetative surveillance for following the main physiological cardio-respiratory variables. The Section 1.4 is dedicated to an example where critical profiles and alarms are built using the big data methodology. This example concerns the detection in a population at risk of pulmonary edema or cardiac failure, in which the part of the vegetative nervous system made of the union at the bulbar level of the respiratory and cardio-moderator centers is represented by a couple of two 2-order differential systems of Liénard type, whose parameter values at rest have been used for building an a priori classification of the cardio-respiratory profiles.

In the Sect. 1.5, the present big data approach is compared with the complex systems one with which it shares many concepts, especially those related to emergent properties due to the combination of a mass of information coming from various redundant sources related to different interaction networks, like the social, genetic, metabolic and physiologic ones, all necessary to implement effectively the predictive medicine of the future.

1.2 The Big Data Approach

1.2.1 Ancestors of the Digitization of Health Data

The first usage of numeric tables in medicine has been proposed by Abū Bakr Muhammad ibn Zakariyyā al-Rāzī (865–935), a Persian polymath and physician called also Muhammad Rhazes [1], who probably wrote the book "The Secret of Secrets" [2, 3], a pseudo-Aristotelian treatise in form of a letter from Aristotle to his student Alexander the Great treating a wide range of topics, including astrology and medicine, and containing two charts from an Arabic copy for determining whether a patient will recover and survive, or remain frail and die based partly on the numerical value of the letters of his name (Fig. 1.2). The old dream of the "Secret of Secrets" will be realized in the future by the predictive medicine, whose aim is to predict for a given patient the destiny of his diseases, even before their phenotypic occurrence.

The founder of algebra in 840, al-Khwârizmî and his numerous disciples like Guillaume d'Auberive (1120–1180) and his Cistercian colleagues Odon de Morimond and Thibaut de Langres [4], or Shaiykh Bahā'ī (1546–1622) [5] knew the science of numbers and the charts of the Secret of Secrets, which is an amazing mix of simple rules, such as arithmetic sequences and some sophisticated rules as the triangle of Tutte [6] (Fig. 1.2), whose properties were surely ignored by Rhazes. It was only in 2010 that the smallest perfect equilateral triangle dissection by equilateral triangles has been exhibited [7], as conjectured by W. T. Tutte. Numbers of the right chart of Fig. 1.2 are inside white external (11, 19, 20) and internal (2, 3) sub-triangles of the Tutte's triangle, or are sums (modulo 30) of the numbers inside these sub-triangles related by red links. Remaining numbers between 1 and 30 belong to the left chart and represent arithmetic series, the first starting at 6 with step 3 (except 12, badly transcribed in 28 on the right chart), the second starting at 5 step 5 (except 20, replaced by 22, a cabalistic number). All these rules constitute a necessary and sufficient condition for belonging to the charts.

The epidemiological approach of the contagious diseases spread using real data is more recent. The first example concerns both social and infectious contagious diseases: St. Anthony Monastery near Grenoble was in charge of watching and curing infectious spreads like the Black Death of 1348 (Fig. 1.3) and social epidemics like the St. Anthony's fire disease (called now ergotism, also known as ergot poisoning). The Antonin congregation was founded in 1095 by Gaston de Valloire, a nobleman of the Dauphiné, and confirmed by the Pope Urban II the same year in thanksgiving for his son's miraculous cure from St. Anthony's fire thanks to the

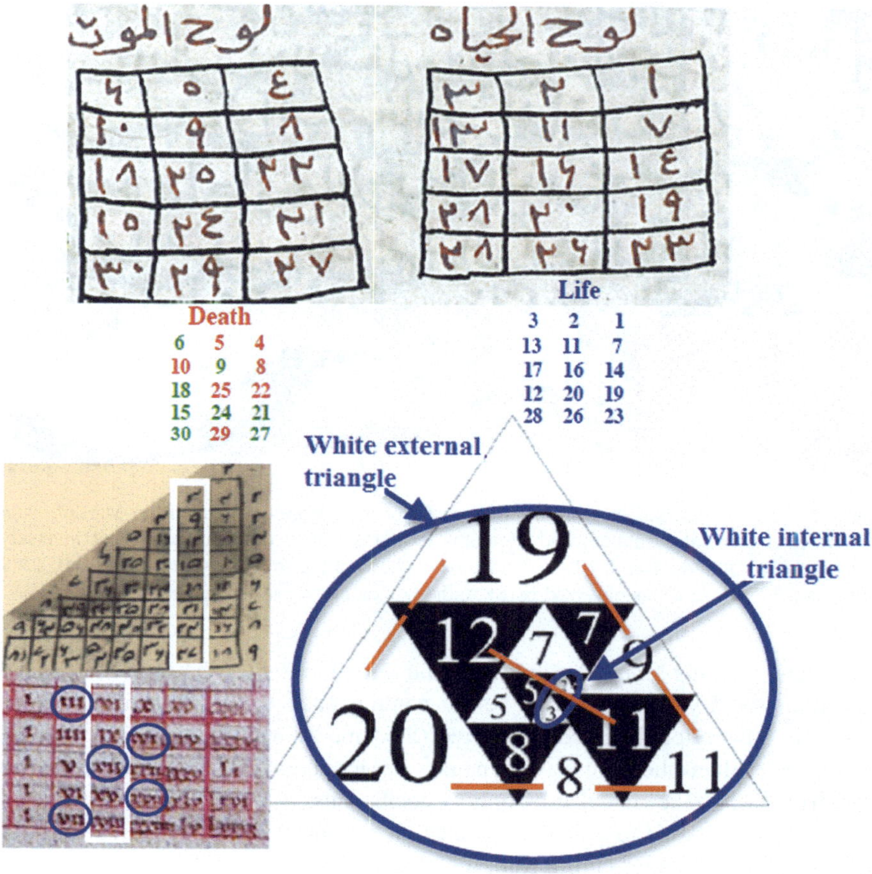

Fig. 1.2 Top: two charts from "The Secret of Secrets" [2, 3], the left (respectively right) chart corresponding to frailty and death (respectively healing and life). Bottom left: arithmetic series by Guillaume d'Auberive (bottom) [4] and Shaiykh Bahā'ī (top) [5]. Bottom right: smallest perfect equilateral Tutte's triangle dissection by small equilateral sub-triangles [6]

relics of Saint Anthony the Great. During the XIVth and XVth centuries, the St Anthony's monastic Order had 640 hospitals and 10,000 brothers in Europe along the roads to Santiago de Compostela and Jerusalem watching plague, leper and ergotism. This monastery is also known for having sheltered two illustrious mathematicians, Jean Borrel (1492–1564), called also Johannes Buteo, an algebraist who refuted the circle quadrature theory [8] and Jean Duchon, inventor of the thin plates splines theory [9–11].

The "Black Death" (or plague) watching by the St. Anthony's Order has been justified by its pandemic character and its dramatic economic and demographic consequences. At least 75 million of persons died from the disease during the Black Death spread. Born in the Caspian area, the wave went through the Mediterranean

Fig. 1.3 Left: simulations showing the front wave of the epidemic spread (from Marseille and Genoa) after 3 months (middle) and 6 months (bottom) with a diffusion coefficient inversely proportional to altitude. Right: Black Death spread showing the front waves of 1348 (red), 1349 (blue) and 1350 (green), as observed by St Anthony's monastery (Left top thumbnail)

sea routes, reaching Marseilles in France and Genoa in Italy at the end of 1347, and spreading for 5 years in Europe before coming back to the Caspian reservoir. A simple Susceptible-Infected-Recovered (SIR) model with viscosity proportional to altitude explains the front wave dynamics observed during the half decade 1347–1352 (Fig. 1.3). The model uses only 3 coefficients, a local diffusion coefficient inversely proportional to altitude, a contagion parameter and a recovering rate (or immunization rate) after curing from the plague disease, and follows the approach proposed at Geneva by Daniel Bernoulli in 1760 [12] and at Paris by Jean Le Rond d'Alembert (the son of the famous Madame du Tencin, close friend of Voltaire living near Grenoble) in 1761 [13]. They first discovered the quadratic equation of epidemic spread. Despite its simplicity, the Black Death model is able to render count qualitatively for the morphology of the front wave and, after time calibration on the first front observed in 1347, for its velocity. Despite the quarantine measure decided after the Black Death epidemics, the plague arrived anew in 1720 at the port of Marseilles on a merchant ship, called the Grand-Saint-Antoine… More than the surveillance of infectious (plague) or social (ergotism) contagious diseases, the St. Anthony Order inaugurated the care of the sick at home. Five centuries after, St. Camillus de Lellis (1550–1614), an Italian priest founded an Order, dedicated to the care of sicks at home. Following Antonin tradition, he created the first dedicated hospices (1584), where he mixed people of all ages and conditions, as precursor of the intergenerational homes. In 1777, this tradition was lost in Dauphiné, when St. Anthony Order was transferred to the Order of Saint John of Jerusalem (currently Order of Malta), but French encyclopedists restored the medieval vision of man as a whole, especially the "Physiocrates", ancestors of the physiologists and early supporters of the health monitoring and education at home, allowed now in the 5P Medicine (Personalized, Preventive, Participative, Predictive and Pluri-expert) by the tools recording and storing health data, from Internet of Things (IoT), hospital data warehouses and big genomic and proteomic repositories.

1.2.2 The Main Big Data Operators: Neural Network Convolution, Renormalization, Gradient Back-Propagation and Simulated Annealing

Neural Network (NN) approach is the core of the modern version of the machine learning called the "deep learning". After a first classical phase where neural networks were used for treating raw data, mimicking some biological operators like the lateral inhibition in retina on the visual way [14, 15] or in cochlear nucleus on the auditory way [16], they were declined on various form for learning tasks [17, 18]. More recently, the invention of the "deep learning" approach has been proposed, often by people like Hinton [19], LeCun [18, 19] and Gallinari [20] already deeply involved in the first classical phase. The main operators on neural networks used in deep learning, i.e., discrete convolution, network renormalization, gradient back-propagation and simulated annealing, have been already proposed either in a continuous mathematical formulation (1952 by L. Schwartz, for the convolution of distributions: "pour régulariser, on convole"), or in a physical framework (1972 by K. G. Wilson, for the group of renormalization), in image processing (1984 by D. and S. Geman, for the simulated annealing) and in recognition tasks (1986 by D. Rumelhart, G. Hinton and R. Williams, for the gradient back-propagation). The main novelty in deep learning is the successive use of these operators for finding hidden relationships between the entities providing data "tsunamis", e.g., caused in medical applications by the multiplication of new medical information sources (from computerized imaging, protein spectroscopy and genome identification tools, to the wireless multi-sensing sources of e-health).

An example of deep learning architecture is the LeNet-5 network by LeCun [19]. Figure 1.4 shows how the sequential use of various operators can extract hidden features from a mass of data, related for example to a genetic network. The initial observation concerns DNA-array data, from which it is decided if a gene is expressing a protein (state 1, corresponding to the protein concentration over a certain threshold) or not (state 0, corresponding to the protein concentration under this threshold). The succession of these binary states in time, e.g. 120 successive records (four blood samples each day during a month) in a big study of 10,000 patients suffering the same chronic disease, thanks to a very high density DNA microarray (e.g., with 10,000 single-nucleotide polymorphism SNP's as probes) gives a part of the trajectory of gene expression configurations with 10^{10} data. The aim of the interpretation of such data is double: (i) identify the final configurations (called attractors) of the dynamics of expression and (ii) confirm the architecture of

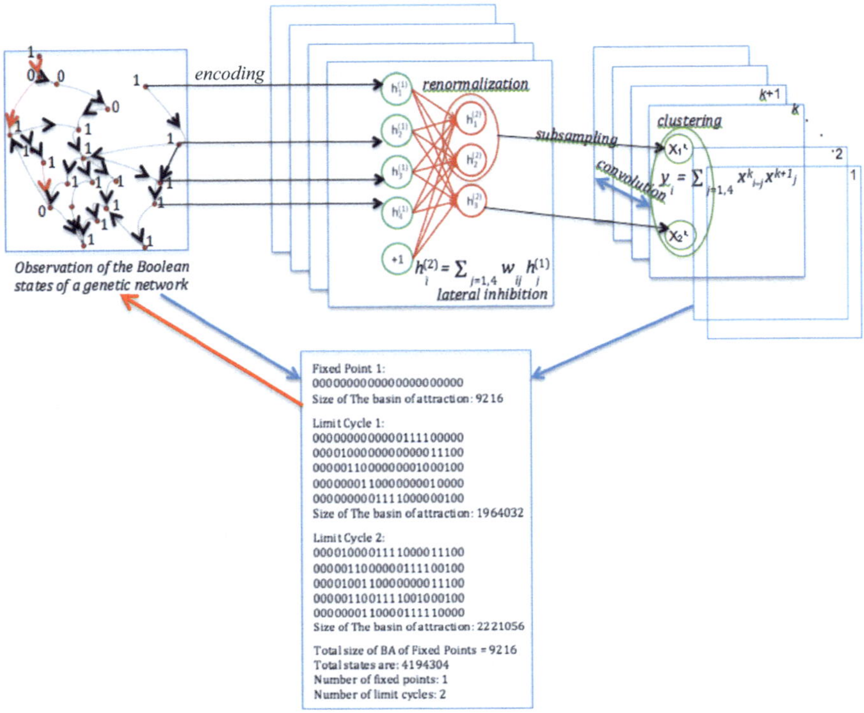

Fig. 1.4 Top left: genetic network whose vertices are genes in state 1 (expressing proteins) or 0 (not expressing). Top middle: stack of several neural networks encoding each a part of the information coming from the observation of the genetic network states, processed by intra-network operators, like lateral inhibition for contrasting and renormalization for averaging. Top right: subsampling and clustering using inter-network operators like convolution. Bottom: clustering in classes corresponding to the final configurations of the expression trajectories (expression attractors)

the genetic network by identifying inhibitions and activations (respectively red and black arrows on Fig. 1.4 Top left). If statistical procedures (mainly based on correlation analysis, in a spatiotemporal Markovian or renewal context [14–17], but with the risk of getting parasitic interactions, due to purely statistical and non causal correlations) are giving too partial information about the interactions between genes and their nature (inhibitory or activatory), the deep learning approach can be used for clustering observed configurations inside classes of states corresponding to the attraction basins of the dynamics of expression, the attraction basin of an attractor being the set of configuration trajectories having final configurations belonging to this attractor (Fig. 1.4 Bottom middle). Of course, if the number of genes is reasonable (about 100, with for example the half located in strong connected components of the interaction graph of the genetic regulatory network) and if all interactions are known in the genetic networks, then a direct procedure of simulation of the Boolean dynamical system underlying the dynamics, allows for obtaining their attractors and attraction basins. If this direct approach (left blue arrow on Fig. 1.4 Bottom left) is impossible (e.g., due to the ignorance on the nature of the interactions), the clustering operator at the end of the deep learning procedure can solve the problem (i) of finding attractors and basins (blue arrow on Fig. 1.4 Bottom right) and after, by solving the inverse problem (ii) (red arrow on Fig. 1.4 Bottom left), give indications for completing the interaction graph of the genetic network.

1.2.3 Big Data Approaches in Health

Many useful big data techniques have been used recently for treating health data, coming from the IoT [21, 22] or from image mining in big medical image repositories or in hospital PACS [23, 24]. The genomic and proteomic data bases (like those located in NCBI—the US National Centre for Biotechnology Information) contain a big amount of information about genes and proteins from numerous different species [25–27]. For example the GtRNA database [27] contains more than 111,000 tRNA sequences in which conserved motifs of length 22 are searched in the tRNA loops, D-loop, anticodon-loop and Tψ-loop [28–30]. Because tRNA sequences possess about 70 bases, this search corresponds to about $2 \times 66\ 10^5 \approx 10^7$ subsequences examination and comparison with reference sequences (Fig. 1.5).

1.2.4 Compressing Before Visualizing

Before the deep learning phase, it is recommended to compress the data, for improving the rapidity of the learning, but also for allowing an easy visualization by the people manipulating the raw information and by the data provider, that is the patient. The compression can be obtained by (i) reducing the dimensionality of the observables (e.g., using Principal Component Analysis or Independent Component Analysis), (ii) discretizing (even binarizing) the data [31–33] (e.g., using fuzzy or Fourier transform, which retain respectively only Boolean or few parameters values) before the steps of classification [34–36] or data mining [35–37].

Apart the medical imaging [23, 24, 38], the main sources of medical information correspond to devices in which a cell, tissue, organ or the entire body is located in a physical field (electromagnetic, gravitational, acoustic, etc.) whose intensity varies abruptly, leaving the cell, tissue, organ or entire body to return to its resting state, this phenomenon being called "relaxation": it is the case for Nuclear Magnetic Resonance (NMR), mass or NMR spectroscopy, Ultra-Sound (US) echography, electrophysiology, etc. In the last case, signals like ECG (Electro-Cardio-Graphy), EEG (Electro-Encephalo-Graphy) or EMG (Electro-Myo-Graphy) can be considered as relaxation signals, and compressed using their decomposition on a base of solutions of various pendulum differential equations.

Number of TGGTA…TTCGA occurrences: 11182
Number of TGGTA… TTCAA occurrences: 7014
Total number of tRNAs: 111 385

TGGTA TTC A 18 196 observed and 111 385 / $10^9 = 0.42 \pm 1$*
>Acaryochloris_marina_MBIC11017_chr.trna26-AlaCGC (603738-603810) Ala (CGC) 73 bp Sc: 77.50
GGGGAATTAGCTCAGCTGGTAGAGTGCTGCGATCGCACCGCAGAGGTCAGGGGTTCGAAT
CCCCTATTCTCCA
>Acaryochloris_marina_MBIC11017_chr.trna43-AlaGGC (4592246-4592318) Ala (GGC) 73 bp Sc: 86.53
GGGGCTATAGCTCAGTTGGTAGAGCACCTGCATGGCATGCAGGGGGTCAGCGGTTCGAAT
CCGCTTAGCTCCA
>Acaryochloris_marina_MBIC11017_chr.trna49-AlaTGC (5637934-5638009) Ala (TGC) 76 bp Sc: 99.24
GGGGGTATAGCTCAGTTGGTAGAGCGCCTGCTTTGCAAGCAGGATGTCAGCGGTTCGAGT
CCGCTTACCTCCACCA
>Acaryochloris_marina_MBIC11017_chr.trna66-AlaTGC (1408891-1408816) Ala (TGC) 76 bp Sc: 99.24
GGGGGTATAGCTCAGTTGGTAGAGCGCCTGCTTTGCAAGCAGGATGTCAGCGGTTCGAGT
CCGCTTACCTCCACCA
>Acaryochloris_marina_MBIC11017_chr.trna8-ArgACG (291610-291683) Arg (ACG) 74 bp Sc: 44.55
GAGAACGTGGTGTAACTGGATAACATCTCAGGCTACGAACTTGAAGATTGAGGGTTCAAG
TCCTTCCGTTCTCG
>Acaryochloris_marina_MBIC11017_chr.trna69-ArgACG (119056-118983) Arg (ACG) 74 bp Sc: 64.84
GGGCTTGTAGCTCAGTGGACTAGAGCACGTGGCTACGGACCACGGTGTCGGGGGTTCGAA
TCCCTCCTAGCCCG
>Acaryochloris_marina_MBIC11017_chr.trna50-ArgCCG (5884305-5884377) Arg (CCG) 73 bp Sc: 75.78
GGGCATGTAGCTCAGTGGATAGAGCATCAGATTCCGGTTCTGAGGGTCGGGGGTTCGAAT
CCCTCCATGCTCG
>Acaryochloris_marina_MBIC11017_chr.trna51-ArgCCT (6020270-6020197) Arg (CCT) 74 bp Sc: 82.76
GGGGCTGTGGCTCAGTTGGATAGAGCAAGCGCCTCCTAAGCGCTAGGTCGGCGGTTCGAG
TCCGCCCAGTCCCG
>Acaryochloris_marina_MBIC11017_chr.trna33-ArgTCG (1924427-1924500) Arg (TCG) 74 bp Sc: 67.64
GGGCTTGTAGCTCAGTGGACTAGAGCACGTGGCTTCGGACCACGGTGTCGGGGGTTCGAA
TCCCTCCTGGCCCG
>Acaryochloris_marina_MBIC11017_chr.trna22-ArgTCT (297423-297498) Arg (TCT) 76 bp Sc: 50.66
GGGAGTGTCGCCTAAGGGATGGGGCACCGATCTTCTAAATCGGTCTGTGTAGGTTCGAAT
CCTACCACTCCTGCCA

Fig. 1.5 Result of a data mining in the NCBI database GtRNA [27] extracting 18,196 occurrences of motifs related to tRNA loops among about 10^7 possibilities (the present list contains only the beginning of the results)

The classical Fourier and wavelet transforms correspond for example to the decomposition of a signal on a family of solutions of differential equations, namely the simple pendulum equation for Fourier transform and the damped pendulum equation, for wavelet transform (Fig. 1.6 bottom). These transforms are well adapted to the case of quasi-symmetric and periodic signals (i.e., whose wave shape has an internal symmetry on a hemi-period, like the sine function), such as the respiratory rhythm (Fig. 1.4 top), for which with few Fourier harmonics, the original shape can be reconstructed with a small controlled error. It is the same for damped quasi-symmetric signals, such as the vibration of the basilar membrane after a short sound stimulus, which is well compressed by wavelet transform.

But relaxation signal like ECG needs a large number N of Fourier or wavelet harmonics to be well compressed (e.g., N = 20 for ECG and Fourier with Signal-to-Noise Ratio (SNR) of about 20 dB using online site http://www.lpsa.swarthmore.edu/Fourier/Series/WhyFS.html#Electrocardiogram). In these cases, a smart compression method consists in decomposing the signal on a family of

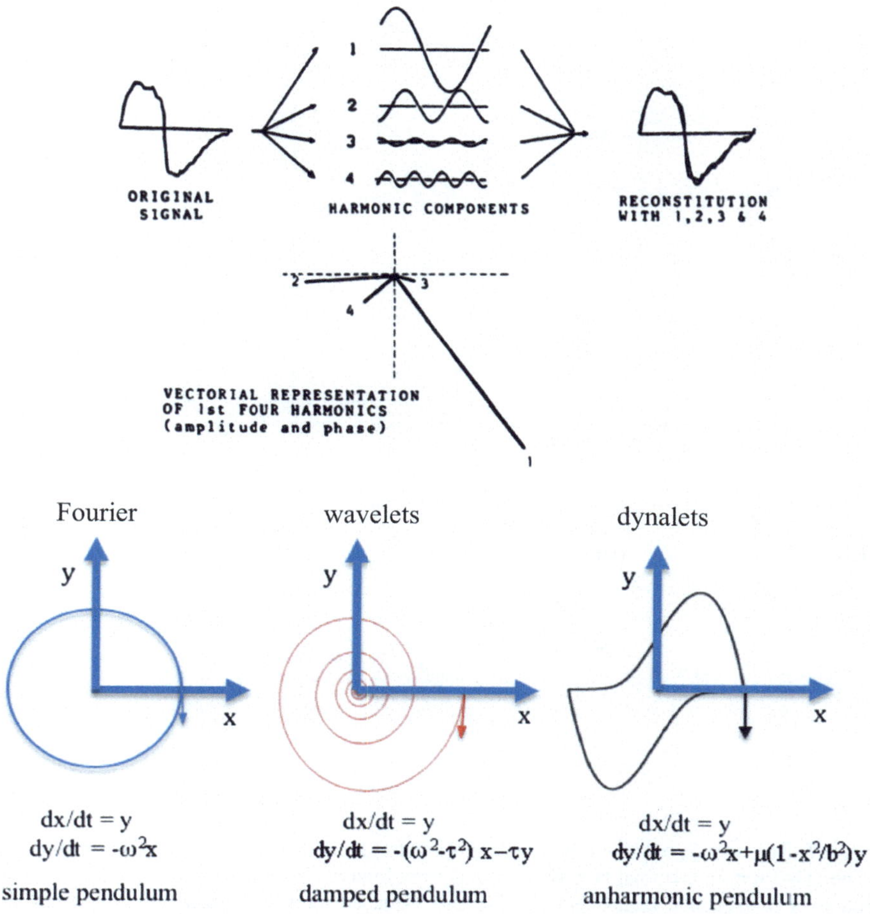

Fig. 1.6 Top: Fourier decomposition of the respiratory wave into one fundamental and three Fourier sub-harmonics, giving a correct reconstitution and allowing for a graphic representation (called respiratory "aster") in the amplitude-phase plane of the signal. Bottom: the 3 pendulum equations corresponding to the 3 transforms, Fourier, wavelets and dynalets [39, 40]

approximate polynomial solutions (called dynalets [39, 40]) of the an-harmonic pendulum equation, proposed first in its analogic version (the "singing arc lamp") in 1900 by W. Du Bois Duddell, then mathematically studied by H. Poincaré in 1908 (who invented first for this equation the term limit-cycle for its unique attractor), and eventually, proposed by B. van der Pol in 1926 to model the heart functioning. Figure 1.6 top shows the decomposition of the respiratory wave into one fundamental and three Fourier sub-harmonics, Fig. 1.6 bottom shows the pendulum equations used in the three decomposition methods, Fourier, wavelets and dynalets, and Fig. 1.7 proposes an example of dynalet decomposition of the cardiac ECG signal.

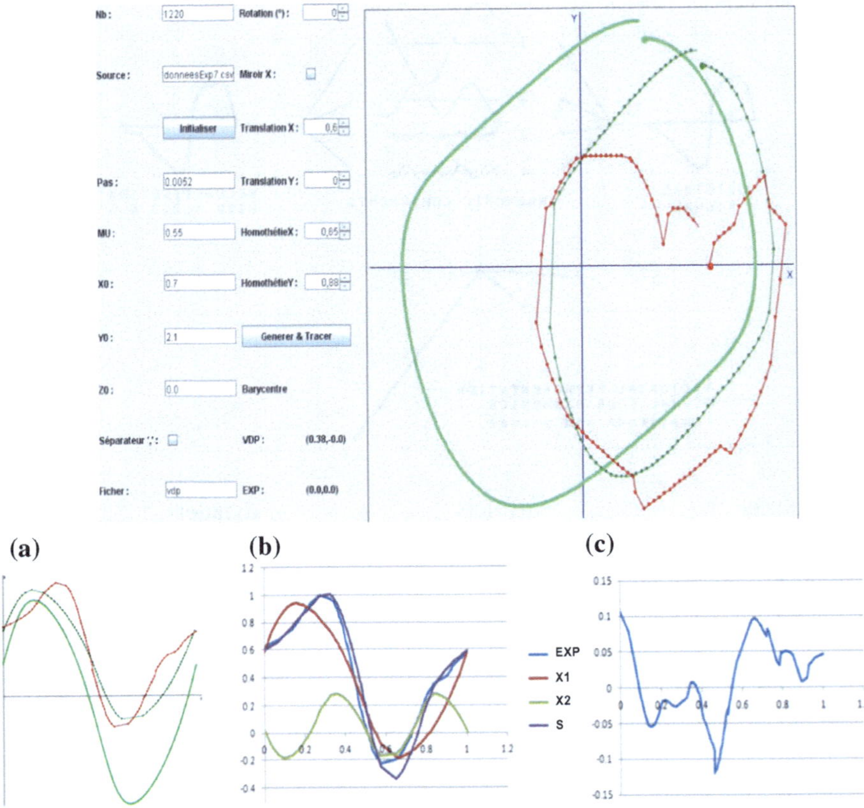

Fig. 1.7 Top: initial position of the van der Pol limit-cycle (in green clear) and ECG signal in the plane xOy (in red), and final fit between van der Pol limit-cycle (in dark green) and ECG signal. Bottom **a** fundamental component (dark green) extracted through dynalet transform from the original ECG signal (in red). Bottom **b** match between the ECG signal (EXP in blue) and the sum S (in violet) of the fundamental component X1 (in red) and the first sub-harmonic (X2 in green), after translation and scaling of the x and y axes, the best fit being searched with a cost function based on the Hausdorff-distance between 100 empirical averages of sampled ECG points and a set of points of same phase extracted from the van der Pol limit-cycle [39, 40]. Bottom **c** signal serving for the calculation of the second sub-harmonics, obtained by subtracting the sum S from the sampled original ECG signal EXP

Another efficient way for compressing the signal is to extract a minimal information corresponding only to the succession of the sense of variations (called monotony signature): +, − or = of the signal, this information being poorer than that coming from rank statistics: observing the rank, we can reconstruct the signature, but inverse is false [41]. Retaining from the signal its monotony signature, consists only in building the succession of + (if intensity of the signal increases), − (if it decreases) or = (if it remains constant) on a succession of windows along the time interval of recording. For example, on Fig. 1.8, the monotony signature of

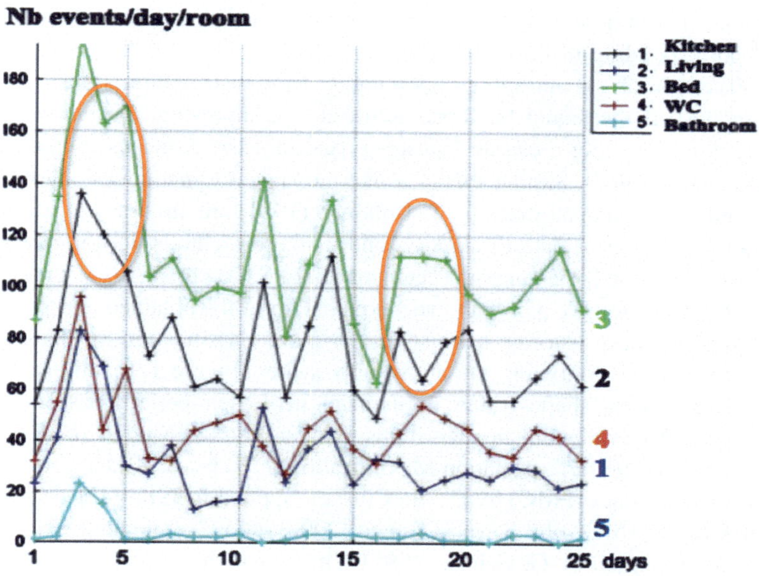

Fig. 1.8 Evolution during successive 25 days of the number of events corresponding to the entrance in different rooms at home during the nychthemeron (day/night 24 h interval). The orange circles emphasize discrepancies between the monotony of the green signal (entrance in bedroom) and the black one (entrance in living)

the green signal (equal to the number of entrance in bedroom of a person at home (suffering Alzheimer's disease with an important cognitive impairment related to a loss of the short-term memory) during a day equals: $+ - + - + - + - + - + + - - + - - - - - + + + -$, signing the probable existence of a compulsive activity such as a pathologic perseveration (often observed in neuro-degenerative diseases).

A simple statistical test of equality of binomial variables shows that there is no significant difference ($p = 0.05$) between the monotony signature of the number of entrance in living (black signal) and of entrance in bedroom (green signal), proving the existence of a correlation between these two signals of entrance. The visualization of the intervals of monotony of various signals (as on Fig. 1.8) by indicating their sense of variation allows for identifying rapidly the discrepancies between signals, and hence, permits to trigger a rapid alarm concerning a pathologic behavior from the personalized actimetry recorded at home.

1.2.5 Clustering Before Interpreting

One of the goals of the big data techniques is to classify, i.e., to dispatch the observations into a family of clusters, in which the mean value of the cluster represents a statistical individual (in general virtual) representative of the cluster population, if it is homogeneous (case of an uni-modal distribution of the cluster values, corresponding in general to a quasi-Gaussian shape for the distribution of the observed variables) and represents nothing if the distribution is heterogeneous (multi-modal). If the signal has been sufficiently compressed, the phase of clustering is rapid by using a classical unsupervised tool like k-means (introduced by J. MacQueen in 1967). Similar methods like the "nuées dynamiques" by E. Diday (1971) and self-organized maps by T. Kohonen (1981) are also easy to use in case of a wide collection of data to classify, and, if these methods do not give results easy to be interpreted, the support vector machine (SVM) by V. N. Vapnik (1991) has the advantage to give a probabilistic explanation of the obtained clustering. The final step of the deep learning can use one or more classification methods (e.g., a supervised method) following the above approaches. Figure 1.9 shows a classification of respiration shapes (after normalization of their period) based on other ideas coming from probabilistic algorithms: Hidden Markov Models (HMM) by R. L. Stratonovich (1960), Expectation Maximization (EM) by A. P. Dempster (1977), Final Markov Classes (FMC) by B. van Cutsem (applied in 1984 [42]), called now Markov CLuster (MCL) by S. van Dongen (2000), and Classification Expectation Maximization (CEM) by G. Celeux (1991). The classification of Fig. 1.9 has been obtained using the FMC classification algorithm [42] and the obtained classes have been proved to be robust after reapplication of the same method on same individuals 4 years after the first recording [43]. The association of a FFT on the respiratory signal and of a Markovian clustering algorithm has permitted to treat rapidly 1000 patients and to show the existence of a respiratory "personality", attested by the permanence of their respiratory profile, even if occur intercurrent diseases or change of habits of life.

Another method of clustering available for big data sample comes from the functional estimation. For example, one device providing spectral signals from protein involved in cancerogenesis is the MALDI-TOF mass spectroscope. Dataset originated from studies of colorectal cancer are available in [44] and correspond to diseased subjects (Fig. 1.10 Top) and subjects of control (Fig. 1.10 Middle) so that the data set contains 112 spectra of length 16,331, among which 64 spectra are those of diseased patients. This particular study contains only a few part of the 250,000 possible protein spectra (20,000 for proteins expressed by the human coding genome and 230,000 for immunoproteins), but the clustering method is the same for more massive cancer data studies.

After denoising the mass spectroscopy signals thanks to wavelet transform, the use of functional Principal Components Analysis (PCA) allows for estimating the local functional likelihood of the curves after randomly partitioning these curves into a learning set of size 80 and a test set of size 32. Based on the optimality criterion for reducing the classification error rate, the number of main components used for diminishing the dimension of the data and constructing a semi-metric permitting the classification, is equal to four, giving an average error rate of $0.06 \pm 0.056^*$ for 100 test specimens [45, 46], which corresponds to a discrimination between the mass spectra better than one made with the classical kernel methods of estimation (average error rate of 0.072 ± 0.033) [47].

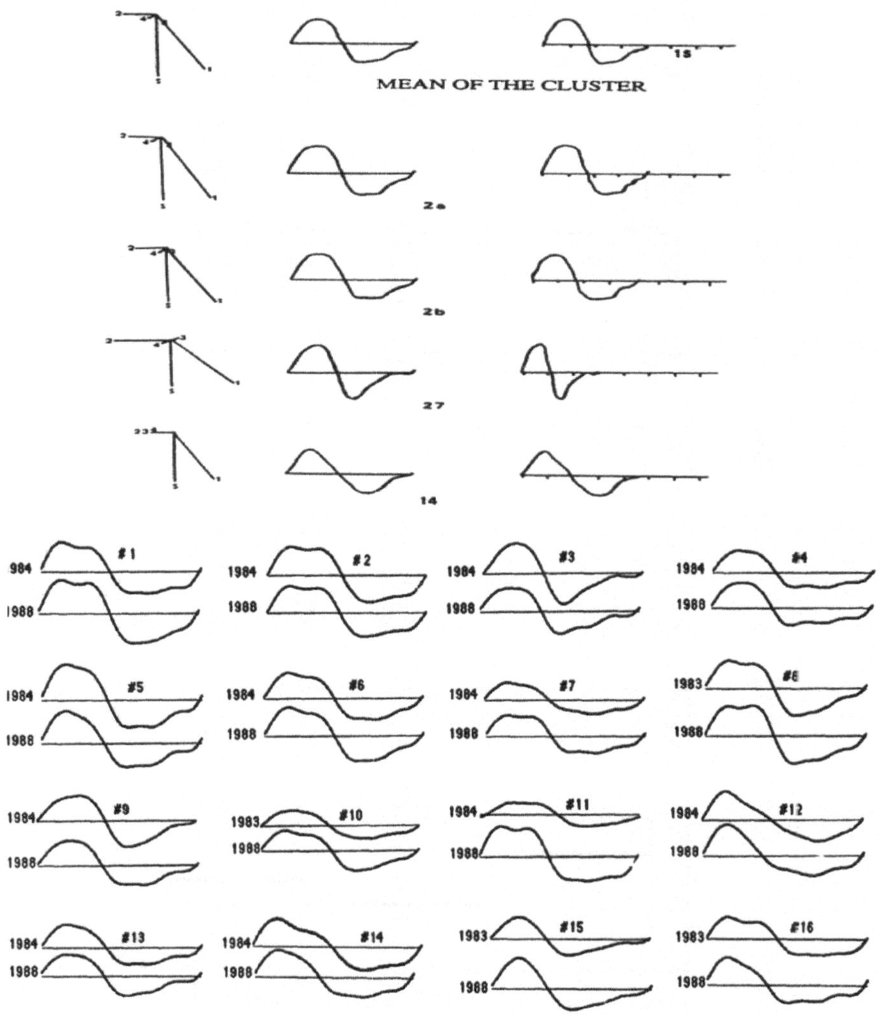

Fig. 1.9 Top: respiratory waves of one class obtained by the final classes algorithm [42]. Bottom: respiratory waves of the same class obtained from records made on the same subjects, but 4 years after [43]. The wave shapes are issued from the harmonic analysis of the airflow signal (approximately 50 respiration cycles per subject), after time normalisation

Figure 1.10 Bottom shows the performance of the functional Principal Components Analysis (PCA) and local functional likelihood approach [46] in terms of Signal-to-Noise Ratio (SNR) compared to the functional Partial Least Squares (PLS) regression [47], showing a disjunction of the spectra between the two classes (cancer and control), which remain confounded in the PLS method. This approach is available for huge data sets and for a large number of classes, offering then an alternative to the traditional estimation methods, which consider the minimization of the least absolute relative error for regression models.

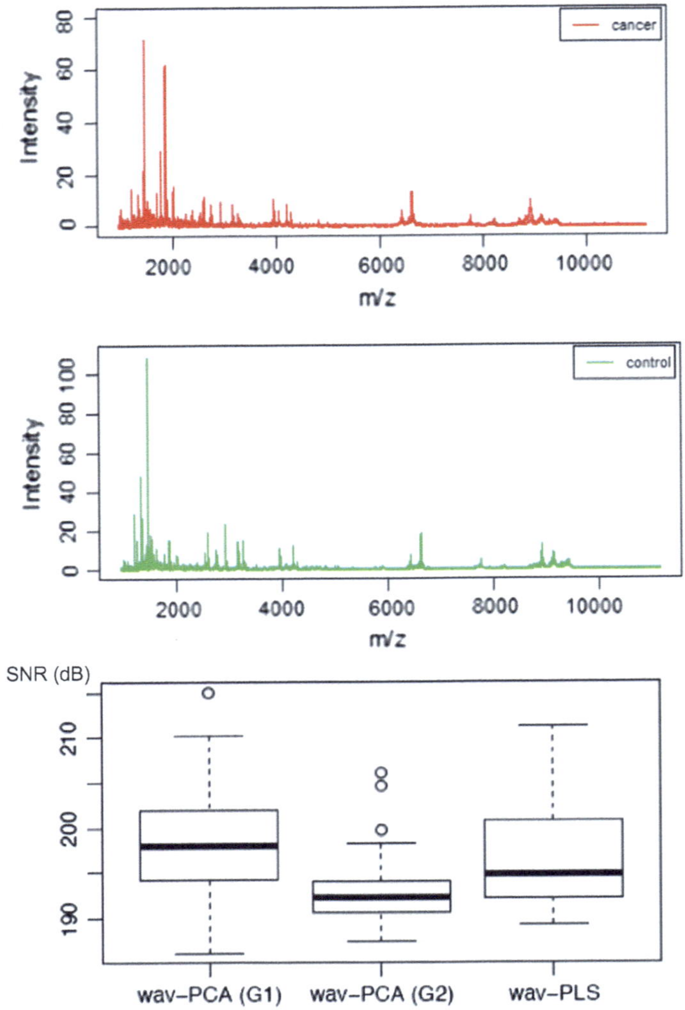

Fig. 1.10 Top and Middle: MALDI-TOF mass spectra from the group G1 of patients suffering colorectal cancer (top) and control group G2 (middle) recorded on the m/z interval of 900–11,160 Da. Bottom: Signal-to-Noise Ratio (dB) compared between the wavelet-PCA and wavelet-PLS methods

We can prove strong uniform consistency, mean squared convergence rate and asymptotic normality of the estimator [45, 46]. That shows the interest to combine quantitative methods of functional non-parametric estimation (i.e., with a minimum of hypotheses on the probabilistic structure of the data) with classical qualitative techniques of non-inferential data analysis, such as the functional Principal Component Analysis.

1.3 The Medical Wireless Sensing

The World Health Organization (WHO) has codified health disability in the International Classification of Functioning (ICF), giving a reference standard for describing the information concerning the well functioning or the disability of a person [48]. The IoT approach in e-Health gives the possibility to build suitable platforms to realize ubiquitous health studies using body and/or environmental sensors and to upload the recorded data to servers to be stored, treated and interpreted for triggering alarms depending on ICF classes at individual or population levels. Mobile Health (m-Health) emerged also recently in healthcare, based on smart phones using several facilities like Bluetooth for interfacing sensors measuring physiological parameters in interoperable environments, such as home monitoring systems for aged-care [49–60], appeared also recently, based not only on sensors and actuators, but also on wearable, implantable or microcapsule devices connected through wired or wireless networks to a service center with diagnosis and therapeutic facilities.

All these devices assess images, sounds and motions through physical parameters (optic, acoustic, thermic, gravitational, electro-magnetic, etc.), as well as vital signs (respiration and cardiac rhythms, blood pressure and oxygenation, body temperature, etc.). The parameters related to daily activity and to sleep are recorded, namely those related to the social life. All the new e-Health, IoT and m-Health devices are involved in the "tsunami" of medical data, to the processing of which the big data techniques combined with recent statistical tools [61–64] seem particularly well adapted.

In the e-Health systems, physical sensors, for example, register the position and the movements of the patient and of their caregivers at home, capturing information about changes of many external physical fields, modified during daily activities by the elderly: thermal field (for early detection of thermal discomfort, premise, in the worst case, of a move toward malignant hyperthermia) using smart vest and bracelets, gravitational field (detecting for example abnormal accelerations of the trunk) using smart vests, bright field using infrared sensors and surveillance cameras, electromagnetic field (used for presence detection) using magnetometers and acoustic field (allowing localization and identification) using microphones, possibly combined for detecting the size of one step, or detecting a fall by using accelerometers or microphones.

e-Health smart home technologies can be classified according to their function [65]:

- Physiological monitoring (measurement of vital signs),
- Functional monitoring (measurement of the activity level, meal intake, etc.) and emergency detection (abnormal or critical situations such as pathologic walk or fall),
- Safety monitoring (occurrence of fire or gas leak), with assistance (such as automatic turning on/off path to bathroom lights when getting out of the bed) and surveillance (alarm triggering in case of abnormal vital signs),

- Security monitoring and assistance to detect and manage human threats such as intruders,
- Social interaction monitoring and assistance like phone calls, visits, coaching and participation in social activities. Assistance includes technologies allowing virtual participation in group activities, video-calls with family and friends, etc.
- Cognitive assistance like automated or self-initiated reminders (lost key locators, medication reminders, etc.) and education technologies (serious games, biofeedback, etc.).

Devices used in healthcare, especially those based on IoT and m-Health technology, could help not only in the care of elderly people but also in the management of medical surveillance of chronic diseases such as cardio-vascular, metabolic, pulmonary, renal or neuro-degenerative diseases. For type 2 diabetics for example, sensors allow for following two potential complications of diabetes: nephropathy and diabetic foot. Several devices are cited in review [66] such as implantable glucose sensor, endo-radio-probe, radio pills, etc. Implantable glucose sensors are based on the 'enzyme electrode' principle, for getting the glucose concentration [67]. Microelectronic devices such as the endo-radio-probe [68] are introduced into the body to record physiological data not otherwise obtainable and to deliver drug in vivo. The principle is to swallow a small transmitter (possibly coupled with a micro-camera) in order to monitor digestive tract parameters (pressure, temperature, pH, etc.). The system consists of a radio frequency (RF) transmitter/receiver and a controller [69].

1.3.1 Smart Homes and Actimetry

Actimetry sensors record positions and movements of a patient and his caregivers at home (Fig. 1.11), capturing information about external physical fields changes (thermal, gravitational, light, electro-magnetic and acoustic) during the daily activities of a person [70–74].

1.3.2 Thermal Sensors

Early detection of thermal discomfort involves capturing multiple information concerning skin temperature and resistivity, room humidity, ambient air flow intensity and direction, etc. The development of micro-power acquisition devices and processing facilities devoted to these data [71] allows for considering the possibility to prevent early abnormalities of thermoregulation caused by a heat wave. The acceptability of sensors allows for capturing the surface temperature on the chest, wrist and ankle, through the assimilation of these sensors to familiar objects within usual clothes, watches or lockets.

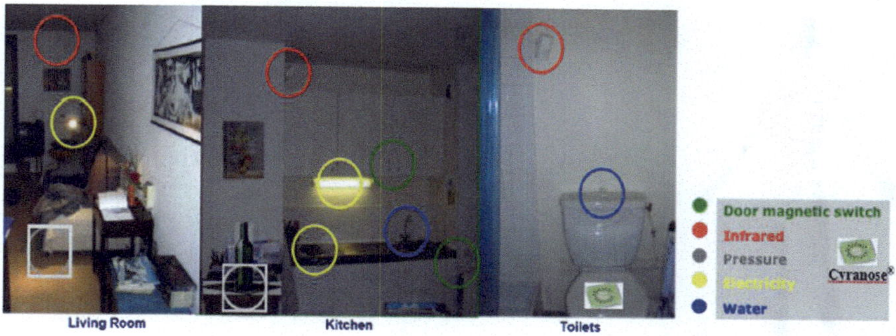

Fig. 1.11 Left: in the living room, pressure sensors under the feet of furniture and web of pressure on the chairs, armchairs, sofas and beds (gray); lighting sensor (yellow) and infrared presence sensor (red). Middle: in the kitchen, pressure sensor under a bottle containing a liquid (gray), water flow sensor at the tap (blue) and electronic switches for opening/closing doors (green). Right: in the toilet, water level sensor (blue) and chemo-sensor or resistivity sensor for urea concentration (green)

- Pyro-sensor away at body temperature

The medicalized antimicrobial lighting sconces by the company Legrand® bring innovative techniques in terms of decision, connectors, and light source. They incorporate a magnetic plug (type Apple®, with which they share the patent, but bigger), a manipulator, standard jacks (2+ earth, for electricity) and FTP (Foiled Twisted Pair) jacks for medical computer equipment [72]. The temperature measurement sensor is a thermopile included in a fixed pyro-sensor placed on the light applied to the top of the bed at the living place, allowing for a double monitoring of presence and hyperthermia.

- Sensor thermal drop

The sensor works by directly measuring thermal infrared radiation in the human environment. It can determine the presence of one or more persons in a room. This information is sent to a microcontroller that transmits data (4 × 4 matrix with 8 temperature levels, see Fig. 1.12) via a USB serial communication to a PC, running the algorithms needed to detect the falling speed [73].

1.3.3 *Gravitational Sensors*

- Fall sensors

A sensorized vest including a fall sensor for detecting abnormal acceleration of the trunk has been developed and patented [60]. This information, cross-checked with actimetric data (pressure sensors on the ground, radar surveillance, infrared sensors, magnetic sensors of door opening, etc.) can trigger a selective information to the

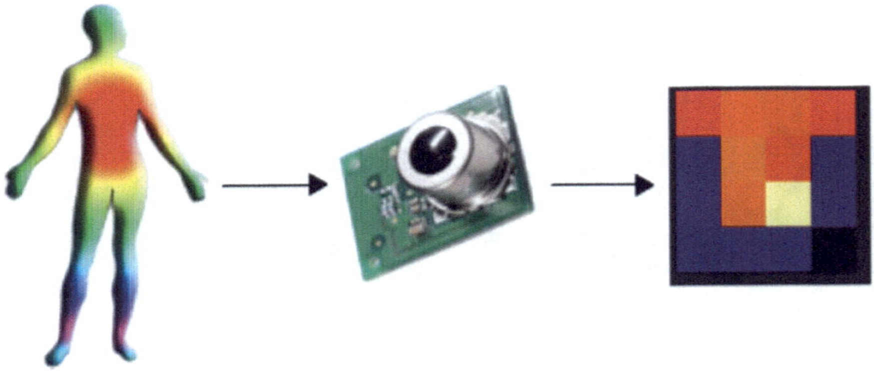

Fig. 1.12 Left: thermal profile of the person. Middle: thermal sensor. Right: 4 × 4 matrix, giving the profile summary with 8 temperature levels

victim of the fall and an emergency intervention if he does not want or cannot react (at least verbally). The acceptability of the fall sensor is strongly linked to its specificity (absence of false positive ensured by the redundancy with environmental sensors).

- Pressure sensors

The principle of the pressure pad is the same as the anti-decubitus actimetric mattresses. In collaboration with the company Léas®, a prototype has been developed, which allows for measuring in real time the pressure on the buttocks of a subject using a wheelchair [75]. This cushion is made of two right and left hemi-cushions (Fig. 1.13). Each half-cushion contains 6 × 12 pressure sensors, each of an area of approximately 1 cm^2. Sensors are non-linearly distributed on the pad, allowing for better resolving zones having ischiatic contacts, where the risk of bedsores are higher. The sensors operate through a semiconductor powder evenly distributed in a polymer shell with elastic properties. The powder acts on the principle of percolation: a pressure force changes its volume and then, its conductivity increases. The current passing through the powder is an increasing function of the pressure exerted. An electronic system connects all sensors, enabling the measurement of an encoded 4-bit electrical potential (corresponding to 16 pressure levels).

The pad is coupled with another prototype, a lingual stimulator named Tongue Display Unit® (TDU), allowing the feedback to the subject [76]. After the seminal work by Kazimierz Noiszewski [77–81], who developed the first device of substitutive vision by tactile stimulation, called Elektroftalm® or “artificial eye” (1897), followed by the Optophone® of Fournier d’Albe (1912) [82], devices have been developed and validated by Samsó Diaz first (1962) and then, by Bach-y-Rita (1969) in a population of deaf animals and blind humans, for which acoustic or visual information captured by microphones or video cameras was transcoded in electro-tactile stimulations of skin or tongue [83–85].

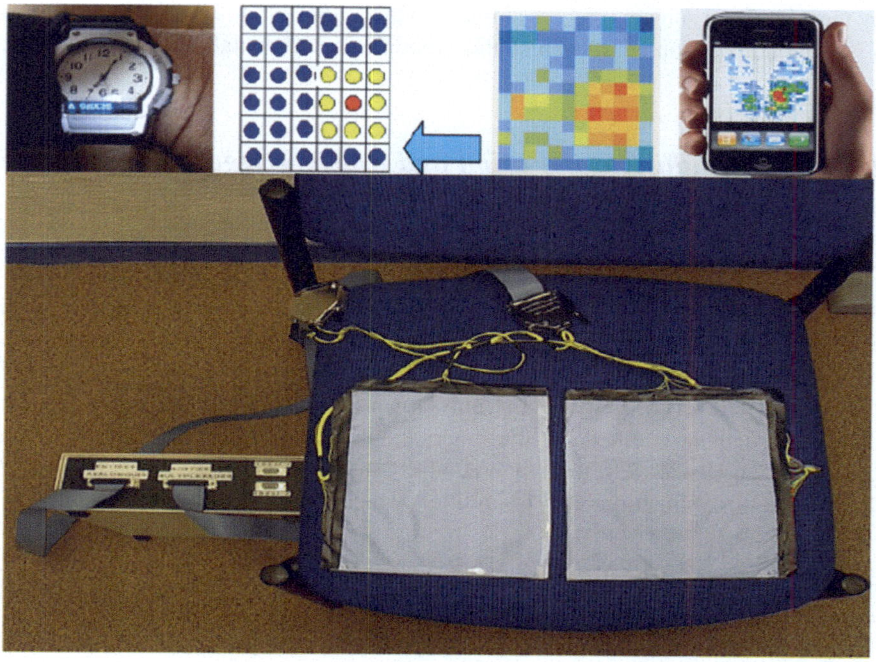

Fig. 1.13 Top left: the watch bottom indicates "go to left" or "escape the right side", where the pressure is the greatest. Top middle: discretization on 3 levels of the information of pressure. Top right: the same pressure information is given on a smartphone. Bottom: pressure recorder cushion on a seat

The tool has been adapted (including wireless) to people at risk of bedsores. The subject can keep the electrodes in contact with his tongue, mouth closed. Saliva having a good conductivity, TDU only requires a voltage of 5–15 V and a current of 0.4–4 mA to stimulate lingual receptors. When an electrode is activated, the subject feels a "tingling" on the surface of his tongue. A subject without spinal cord injury can thus perceive and interpret information provided by the TDU electrodes placed on his tongue and adopt a postural attitude adapted to the information coded in false color like scale of electrical stimuli (Fig. 1.12) in order to limit the over-pressure zones. The acceptability of the transmission of the dermal pressure information requires the incorporation of stimulation electrodes in an artificial palate, which may be that of the dental prosthesis.

The pressure information can also be delivered as a vibration signal at the posterior surface of the upper incisors causing a correction of posture, which can be unconscious in healthy subjects, like in the case of the dermal pressure messages from the sole of the feet [86, 87]. The noninvasive nature of the capture and restitution has been demonstrated, and further studies could show that after a learning phase, the recruitment of the cortical areas of projection of the tongue sensitivity, would definitely prove the phenomenon of substitution and habituation by the stimulated subject (process similar to the gradual habituation of the tongue to a new mouth environment, even slightly modified, after dental work by a dentist).

1.3.4 The Multi-sensor Fusion and Alarm Triggering

- *Generalities on the multi-sensor fusion*

Alarms generated by a monitoring system at the living place (e.g., at home or in a residency for senior) do not directly concern the diagnosis of a disease, as could a medical expert system, giving a code related to the International Classification of Diseases (ICD/DCI) of the World Health Organization (WHO) [48], but indicate a functional deficit, codified in the International Classification of Functioning, Disability and Health (ICF/ICF) of WHO [48], in order to provide a standard in the description of dysfunctions and disabilities, which belong to an ontology of common concepts that concerns:

- organic functions and anatomical structures of individuals
- activities of individuals in areas of social life in which they participate
- personal and environmental factors that influence this participation.

ICF does not classify people, but describes them through multiple hierarchical as well as non-hierarchical classifications, consisting of categories qualitatively coded for describing cognitive and/or sensory-motor impairments, decrease of the participation to the social life, etc. ICF information can trigger an alarm, if the person enters a new set of critical classes requiring attention (immediate or possibly delayed).

- *Multi-sensor integration: the example of detecting the fall*

The function of the fall surveillance system is to detect a person who loses his balance and falls on the floor, compromising his safety and physical and/or intellectual integrity. The operating principle is based on a sound or an infrared signal passing over a critical threshold and generating an alert. The infrared signal comes from a sensor capable of delivering a local temperature pattern in a 4×4 pixel matrix, within a sensing range of about 9 m^2. This information is sent to a microcontroller that transmits data via a USB serial communication to a computer of the ODROID family, integrating these data with the sound signals from a microphone or those of a pressure pad, and executing the algorithms required for detecting the fall as a critical negative acceleration, the common feature of fall sensors [88–94].

The system can be installed for example in the bedroom, the bathroom and the toilet, frequent places of nocturnal fall. After an adjusting phase, the system automatically sends a SMS or email alert to the medico-social environment of the monitored person. The fusion algorithm combines information from infrared sensors with that coming from CMT (Microphone Technology Coincidence) microphones and pressure sensors [94]. The specific constraints of the system are:

- the measured precision is very different for different sensors: a room or a large area of the room for infrared sensors, 15° azimuth for CMT microphones and at most 1/2 m^2 for pressure sensitive tiles or pads,
- the presence information from different sensors is not always guaranteed: CMT will be effective only in case of fall with noise, cry or call by the monitored person (in this case, CMT localization signal has to be filtered by the information coming from an associated smart sound system, which detects and interprets the speech or the abnormal silence)
- the detection has to be performed in real time and to be reliable even the person is not alone.

- **Example of aid to the determination of the plantar balance**

A podiatrist wishing to assess the sheet balance before care and massage and in case of feet ulcers in a complicated type 2 diabetic, can use an important information on how abnormal is the walk of his patient, if it tends to antalgic gait, corresponding to a limp which allows for avoiding pain on weight-bearing structures, e.g., an equinus gait, characterized by a tiptoe walking, to escape a pain of a heel ulcer.

A smart sock (respectively sole), developed by the company Texisense® [86] (respectively FeetMe® [87]) tracks the patient's movements when walking and a bone imaging (Fig. 1.14), recorded by a portable ultrasonic device (e.g. from GE

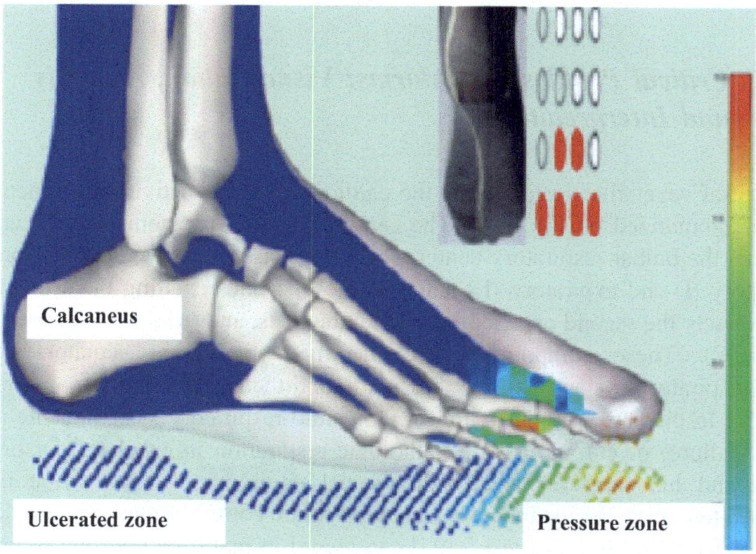

Fig. 1.14 The area of antalgic gait on the forefoot to escape ulcerative heel pain is recorded by the smart sock of Texisense®, which captures the pressure map under the feet (area in red on thumbnail image)

Healthcare®), serves to mapping a default of ossification of the calcaneus (bone whose resorption/reconstruction by osteoclasts and osteoblasts has a turnover/remodeling phase of about 17 months), when it is facing a heel ulcer. The bone tissue is indeed constantly renewed, process disrupted in diabetes, by a loss of osteocalcin secretion and plantar pressure stimulation (due to the antalgic avoidance of the ulcerated parts of the feet).

- **A necessary data fusion**

All the data collected from dedicated fall sensors and from environmental actimetric sensors need to be analyzed separately and then, merged by using a fusion algorithm, in order to be sent to a big data classifier, able to detect the entrance in a zone at risk. Figure 1.15 summarizes the different steps of the fusion procedure and alarm triggering. The scoring using ICF classification gives a scale on which the assignment to a class of risk is made. The alarm information is transmitted both to the patient (if he has still the cognitive and sensory-motor capacity to interpret it and to react) and to his caregivers, as well as to his medico-social environment, triggering for example an immediate help at home. An effective fusion of physiological variables coming from body sensors and environmental variables—more reliable and accurate than the natural one's—coming from external sensors (on walls or furniture) is for example crucial to enable individuals suffering spinal cord injury or having a somatosensory loss in the feet (e.g., coming from a peripheral neuropathy, consequence of a type II diabetes) to avoid a local excess of pressure at the skin/seat interface leading them to adapt their posture in order to prevent falls and/or ulcers.

1.3.5 Critical Profiles and Alarms: Visualization, Analysis and Interpretation

The central vegetative system rules the cardio-respiratory activity. Its functioning can be summarized on Fig. 1.16. The central vegetative system has two components; (i) the bulbar respiratory center which consists of two types of neurons, the inspiratory (I) and expiratory (E) neurons, the first ones exciting the second ones and inversely the second ones inhibiting the first ones, and (ii) the cardio-moderator center (C). These components rule the main peripheral actuators of the cardio-respiratory system, i.e., the diaphragm and the heart controlled by the sinus node (S). In the case of dysfunction of the cardio-respiratory system, acute (like a cardiac failure) or chronic (like a pathologic respiration in an obese person), the patient and his caregiver and medico-social environment have to be rapidly informed to correct the dysfunction and avoid future complications. Several devices like the smart clothing Visuresp® [95] can record in real time both the respiratory and the cardiac rhythm and send the information to a center monitoring many patients like the Kaplan center of cardio-respiratory rehabilitation at Valparaiso in which an innovative remote rehabilitation program for the cardiovascular patients

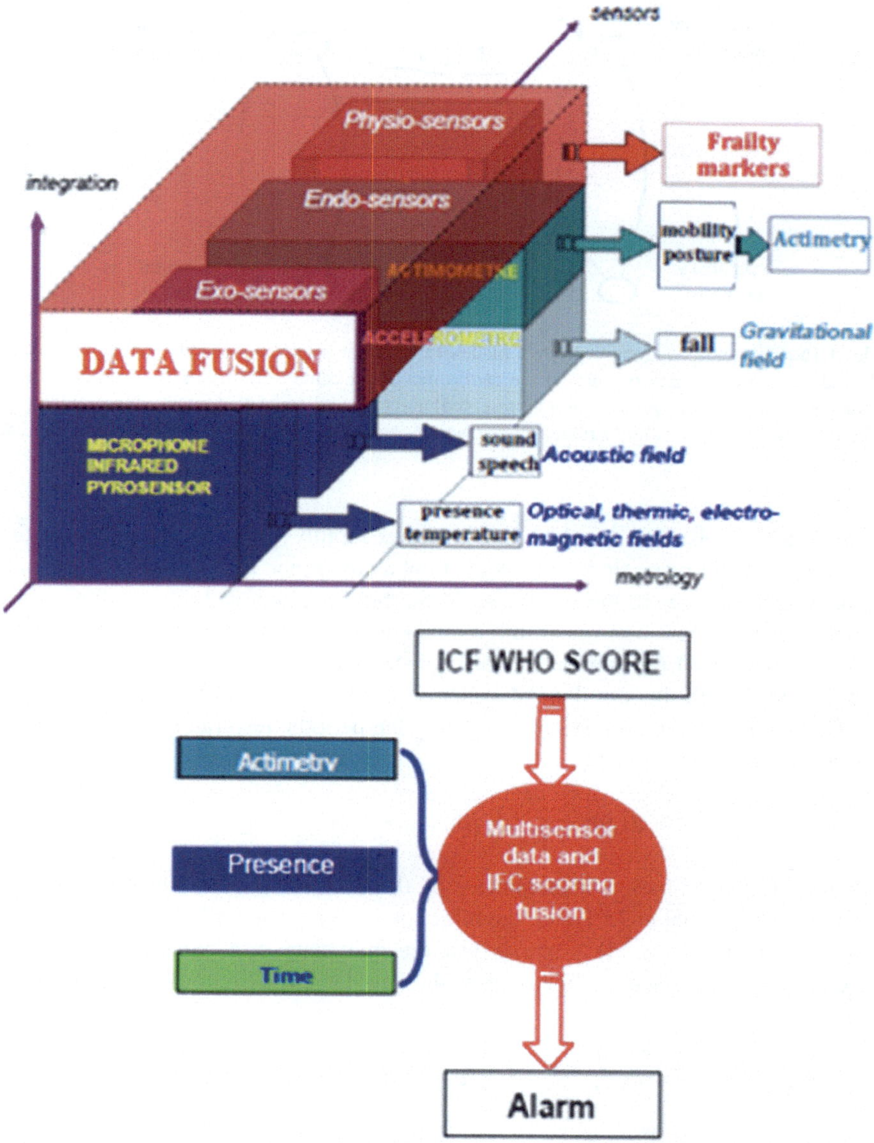

Fig. 1.15 Top: fusion of data coming from various types of endo- and exo-sensors detecting acoustic, gravitational, thermal, optical, as well as electromagnetic fields. Bottom: fusion of data from multiple sources such as presence sensors needed to calculate a profile from the IFC score and then, trigger an alarm if there exists a life-threatening risk, like a fall or a hyperthermia

of Dr. Gustavo Fricke Hospital has been developed by the University of Valparaiso and the Viña del Mar Quillota Healthcare Services in collaboration with the Jorge Kaplan Foundation, thanks to an award of the Innovation Fund for Competitiveness (FIC) of the Regional Government of Valparaiso [96].

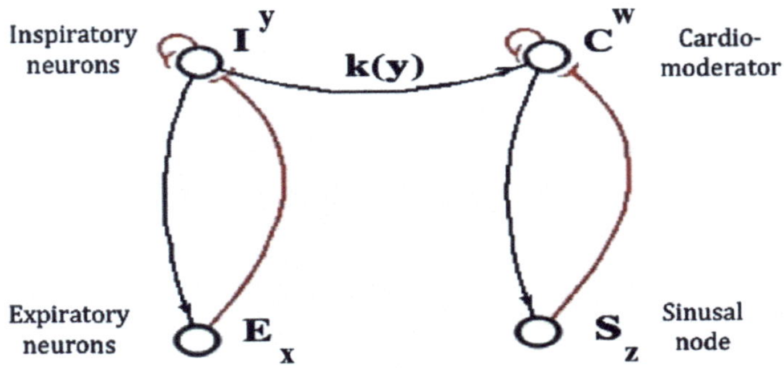

Fig. 1.16 The central vegetative system made of the bulbar respiratory center with its inspiratory (I) and expiratory (E) neurons and of the cardio-moderator center (C), ruling the main peripheral actuators, namely the diaphragm (not represented) and the heart controlled by the sinusal node (S). The four variables x, y, w and z represent respectively the electrical activity of the four centers E, I, C and S

In order to visualize, analyze and interpret rapidly the data coming from the Visuresp® cardio-respiratory sensors (airflow recorder and ECG as on Fig. 1.17), it is necessary to enter in the last phase of the big data approach, the mathematical modeling, using here non-linear differential equations to account for the relaxation behavior mainly of the cardiac, but also of the respiratory components, for which the modeling with the van der Pol system has been chosen [40].

The van Pol differential system modeling the periodic respiratory activity can be written as follows:

$$dx/dt = y, \quad dy/dt = -x + \varepsilon(1 - x^2)y \tag{1.1}$$

where ε is the an-harmonic parameter of the oscillator, with a proper period τ approximately equal to $2\pi/I$, near the bifurcation of the van der Pol system at $\varepsilon = 0$, where I equals $(2 - \varepsilon^2/4)^{1/2}$ and is just the imaginary part common to the two eigenvalues of the Jacobian matrix J of the van der Pol system (1.1), given by:

$$J = \begin{pmatrix} 0 & 1 \\ -1 - 2\varepsilon xy & \varepsilon(1 - x^2) \end{pmatrix}$$

The van der Pol differential system modeling the rhythmic cardiac activity can be written as follows:

$$dz/dt = w, \quad dw/dt = -z + \eta(1 - z^2)w + k(y)y \tag{1.2}$$

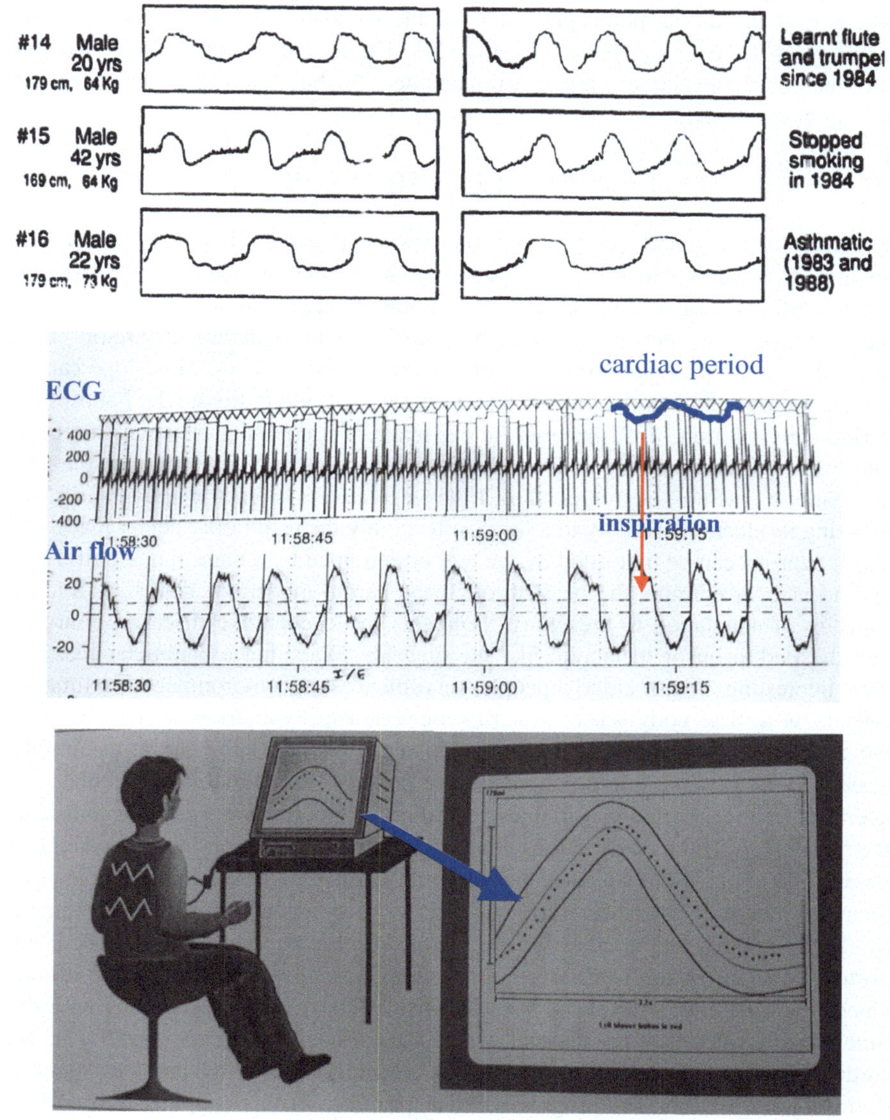

Fig. 1.17 Top: individual original airflow recordings coming from two studies separated by four years, with indication of some physical characteristics observed in the first study, showing the conservation of a respiratory profile, which defines the signature of the respiratory personality [43]. Middle: evolution of the instantaneous cardiac period, which is anti-correlated with the inspiration duration. Bottom: biofeedback rehabilitation using the smart shirt VisuResp®, of a patient suffering a respiratory disease such as a broncho-obstructive pathology

where η is the an-harmonic parameter of the oscillator and k(y) represents the intensity of the coupling between I and CM. The period of the cardiac oscillator entrained by the respiratory one is estimated near the bifurcation of the van der Pol system at $\eta = 0$, by:

$$T = 2\pi/(1 - \eta^2(1 - (k(y)y)^2)^2/4)^{1/2}.$$

The values of ε and η depend on the experimental periods of the cardiac and respiratory systems and then, k(y) comes from the measurement of the instantaneous cardiac period T, i.e., the inter-beats duration, after estimation of the slope of the regression line between T and the actual instant θ in the inspiratory time interval, where the cardiac beat of period T occurs. The value of this slope can be calculated from the correlation coefficient r between T and θ. Figure 1.17 shows the periodical evolution of T, proving the existence of a coupling between the respiratory and cardiac oscillators. The integrity of the system coupling the respiratory and cardiac rhythmic activities allows the vegetative system for anticipating and adapting its electrical activity to a future effort: only the respiratory center made of I and E neurons can be first ruled by the will or entrained by a muscular activity. In a second step, it entrains the heart through the cardio-moderator center C. Such a capacity of adaptation to the effort disappears in degenerative diseases, neuronal like the Parkinson or metabolic like the diabetes. Watching a parameter like r is then interesting in the elderly people surveillance and environmental actimetric sensors as well as body sensors such as the Visuresp® which allow for observing the emergence and controlling the vegetative regulation of the cardio-respiratory system. If we observe a large population of patients, it is possible to calculate the averages of the vegetative variables recorded by the cardio-respiratory and actimetric sensors. These observations contribute to the constitution of a physiological knowledge about the coupled respiratory and cardiac controls. If the whole nycthemeral (day-night) cardio-respiratory activity of a patient equipped for example by the Visuresp® and/or other specific cardio-respiratory sensors is recorded, the normal behavior of the cardio-respiratory coupling shows that the cardiac instantaneous period is anti-correlated with the inspiration duration, as well as with the time in the inspiratory time interval, where the cardiac beat occurs (Fig. 1.17): the cardiac rhythm is accelerating during the inspiration, and the heart decelerates during the expiration. The progressive disappearance of this coupling during neurodegenerative pathologies like Parkinson's or Alzheimer's diseases allows for diagnosing early the entrance in the chronic dysfunctions of the cardio-respiratory system related to these diseases.

In order to restore an healthy functioning, for example in a patient suffering a broncho-obstructive pathology, a biofeedback rehabilitation is possible, which would advice the patient to follow on a screen an ideal respiratory rhythm calculated from his respiratory signature [43], compared to the normal cardiorespiratory behavior of the class to which he belongs, class provided by the big data procedure of clustering. The assignment to a class of patients at risk can be made also by considering the social network surrounding the patient, in which there are normal

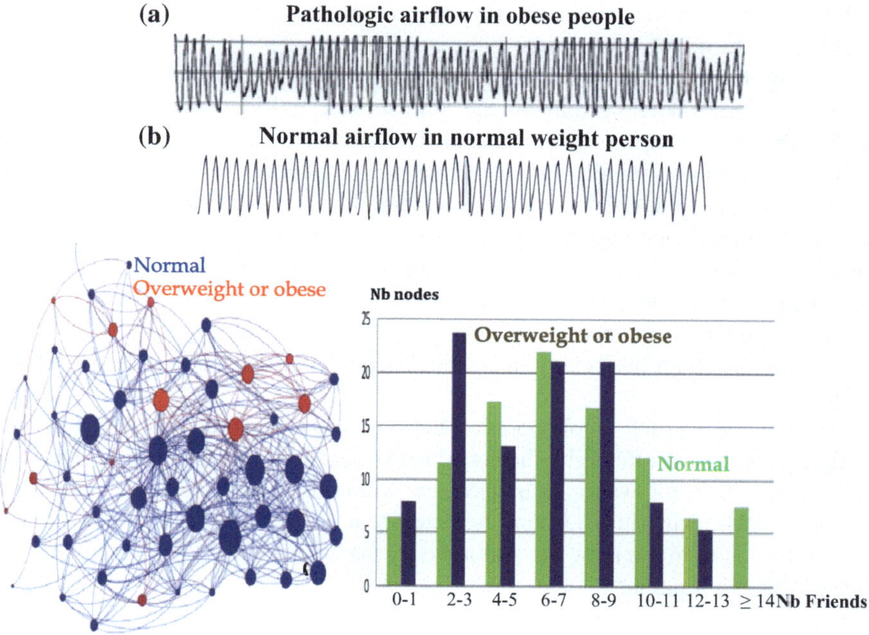

Fig. 1.18 Top: **a** periodic breathing of an obese patient, showing “crescendo-decrescendo” pathologic changes in the respiratory rhythm without central apneas, **b** air flow rate of a normal weight people without any cardio-respiratory disease. Bottom left: social network with normal weight (in blue) and overweight or obese (in red) people, the social links representing the friendship network and node area being proportional to the number of such links. Bottom right: distribution of the number of friendship links inside the normal weight (in green) and overweight/obese (in violet) populations, showing the existence of two maxima (bimodality) in the distribution of the number of friends in the overweight/obese population

people to imitate (in particular in their alimentation or way of life habits), and pathologic patients to avoid, especially those having a negative influence on their human environment (Fig. 1.18). Depending on the state of a patient and also on his environment in the social network to which he belongs, it is possible to personalize a preventive and therapeutic education program [97], whose aim is to globally reduce the occurrence of diseases in the social network (cf. Sect. 1.4).

1.4 Discussion

The complex system approach aims to extract emergent properties from the observation and modelling of the entities in interaction, which provide large spatio-temporal data sets corresponding in general to robust complex systems well regulated in the framework of the homeostasis, notion which characterizes the stability of the genetic, metabolic and physiologic networks that regulate all the interactions inside a living system [98].

The big data approach would converge with the complex systems one, because the explanatory modelling remains the ultimate step in the analysis of mass data, and the theory of complex systems makes it possible to define a minima models that can account for the variability of data with a small number of parameters and also to extract emerging properties, serving as a refutation tool for making these models more realistic and therefore more predictive, especially in medical applications. The future strategy in medical data processing would be thus constituted by a successive use of methods coming from big data tools and others related to the theory of complex systems, combined in order to face all the challenges posed by the mass information in health (especially in e-health), from its visualization to its modelling [99].

An example of such a collaboration between the two paradigms, big data and complex systems, is offered by the social networks. The complex system approach of social networks allows for defining a new notion of centrality, called the entropy centrality [91], taking into account the heterogeneity of the states of the neighbors of a given node i in the network, and not only the connectivity of its neighborhood in the interaction graph:

$$C_i^{entropy} = -\sum_{k=1,\ldots,s} v_k \mathrm{Log} v_k,$$

where v_k denotes the kth frequency among the s frequencies of the histogram of the state values observed in the neighborhood of i, denoted V_i, i.e., the set of the nodes linked to the node i.

This new notion of centrality is for example more useful than the previous ones to detect which obese patient is a good candidate to be educated through personalized therapeutic advices given at home (depending on the observation at home of his actimetric and metabolic data) in order to transform his pathologic state in a normal state, because this candidate can influence efficiently a heterogeneous environment, pushing his neighbors to recover their weight normality. An illustration of this fact is done in the example of Fig. 1.19, which represents the result of the observation of a large social networks with obese, overweight and normal weight persons in familial or friendly relationships, having positive or negative influences on their neighbors. This influence, if it comes from an obese and if it is positive, can push his normal friends to adopt his bad way of life and nutrition habits, hence causing their passage from the normal weight state to the overweight one. In order to modify this bad influence in the social network, it is possible to detect the critical obese people having in their environment a sufficient number of normal or overweight people susceptible to be influenced.

A big data approach in a huge social network is capable to detect rapidly the critical "hubs": there exist classical notions of centrality (degree, betweenness, closeness and eigenvector centralities, well known in graph theory) allowing for identifying the critical nodes susceptible to become good target of a personalized

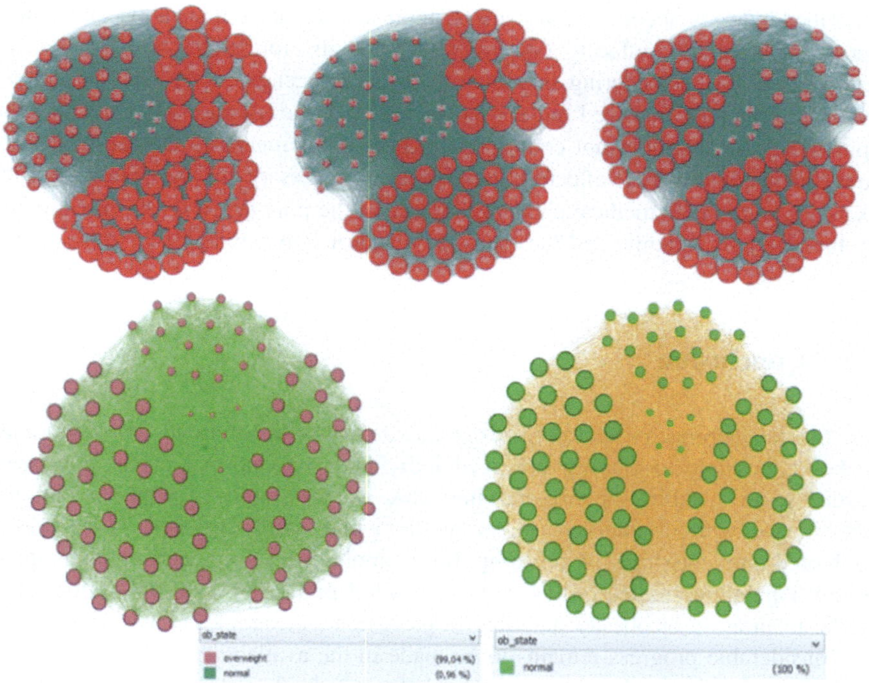

Fig. 1.19 Top: representation of the interaction graph of a part of a huge social network containing obese (red), overweight (pink) and normal weight (green) individuals in interaction (family members or friends). The nodes size corresponds to the in-degree (left), eigenvector (middle) and total degree (right) centralities. Bottom: threshold for the success of a therapeutic education of the N individuals having the largest entropy centrality: after stabilizing the contagion dynamics, all individuals are overweight (left) if N = 20, and all individuals are normal (right) if N = 21, this last number being the critical threshold ensuring the success of the therapeutic education

therapeutic education, and the new notion of entropy centrality has been tested with respect to the previous ones.

By comparing the results obtained thanks to a personalized education leading obese nodes to recover the normal weight state, we observe in the example of Fig. 1.19 that after educating only the 21 individuals having the most important entropy centrality (Fig. 1.19 Bottom right), all the population is going back to the normal weight state, but that needs 68 individuals with the in-degree centrality, and 85 individuals with the eigenvector centrality (Fig. 1.19 Top and Midde left). Then, the best public health policy against the obesity pandemics consists in using the notion of entropy centrality to select the targets of the therapeutic education.

The example of Fig. 1.19 shows the necessity for the big data and complex system approaches to collaborate, because the big data techniques are very useful in the first steps of data processing (compression, visualization and interpretation), but have to be completed in the explanatory phase of modeling, that can in general be described in terms of complex system, for representing interactions between healthy persons and patients suffering a disease [100], and also for dealing with the disease pathogenesis, by considering the network of all the genes, metabolites and environmental factors involved in the occurrence of the disease [101]. The two approaches are therefore not contradictory, but complementary, and must demonstrate in the future their effective collaboration, necessary for the resolution, for example, of the new medical challenges posed at the population level, for example by the obesity pandemic and the return of infectious epidemics [91, 102].

1.5 Conclusion

The big data approach used to organize data sets from e-health multi-sensor systems makes it possible to classify observed individuals into normal behavior classes requiring no specific support and into risk classes requiring either a chronic surveillance without intervention, or an immediate care triggered by an alarm calibrated on a classification resulting, for example, from deep learning techniques in the population studied (which may be several million in size, for example for cardio-respiratory or obesity pathologies presented as examples in this chapter).

Considerable progress remains to be made in the availability and interpretation of data, especially for the patient and his helpers at his/her place of life. The rapid and informative restitution of a summary of patient data is more than an interesting scientific challenge, but also a legal obligation in certain countries (such as France), an ethical necessity and an effective means of empowering people, then more inclined to follow the preventive and therapeutic recommendations made by the health services (doctors and all para-medical professionals) concerning their illness and its possible complications.

Acknowledgements We acknowledge the Projects PHC Maghreb SCIM (Systèmes Complexes et Ingénierie Médicale), ANR e-swallhome and H3ABioNet for their financial support.

References

1. Rhazes M (about 930) (1964) Manfe' al Aghzie va Mazareha (Benefits of food and its harmfulness). Taban Press, Tehran
2. Rhazes M (about 925) (1497) De regimine sanitatis, or Epistula Alexandro de dieta servanda. Juan de Sevilla Hispanus (Transl.) (about 1145). Antoine Vérard, Paris
3. Bacon R (1280) (1920) Secretum secretorum cum Glossis et Notulis. In: Steele R (ed) Opera hactenus inedita Rogeri Baconsi. Clarendon Press, Oxford, pp 287–313
4. Albæripæ W (about 1160) (1165) Tractatus Numerorum a ternario usque ad duodenarium, ad Thomam monachum, vol 969. Médiathèque de Troyes incunable, Troyes, pp 195–196
5. Shaiykh Bahā'ī, M (about 1610) (1843) Kholasat al-Hesab. The summa of arithmetic. In: Nesselmann GHF (trans.) Essenz der Rechenkunst von Mohammed Beha-eddi ben al-Hosain, al-Khulasa fil hisab. G. Reimer, Berlin

6. Tutte WT (1948) The dissection of equilateral triangles into equilateral triangles. Proc Cambridge Phil Soc 44:463–482
7. Drápal A, Hämäläinen C (2010) An enumeration of equilateral triangle dissections. Discrete Appl Math 158:1479–1495
8. Buteo J (1554) Opera geometrica. T. Bertellus, Lyon
9. Duchon J (1976) Fonctions spline et espérances conditionnelles de champs gaussiens. Annales Scientifiques de l'Université de Clermont-Ferrand 14:19–27
10. Duchon J (1976) Interpolation de fonctions de deux variables suivant le principe de la flexion des plaques minces. RAIRO (Analyse Numérique) 10:5–12
11. Duchon J (1977) Splines minimizing rotation-invariant. Lect Notes Maths 571:85–100
12. Bernoulli D (1760) Essai d'une nouvelle analyse de la mortalité causée par la petite vérole, et des avantages de l'inoculation pour la prévenir. Royal Academy of Sciences, Paris
13. d'Alembert J (1761) Onzième Mémoire: sur l'application du calcul des probabilités à l'inoculation de la petite vérole; notes sur le mémoire précédent; théorie mathématique de l'inoculation In Opuscules mathématiques. David, Paris
14. Demongeot J, Fouquet Y, Tayyab M, Vuillerme N (2009) Understanding physiological & degenerative natural vision mechanisms to define robust contrast and segmentation operators. PLoS ONE 4:e6010
15. Demongeot J, Mattes J (1995) Neural networks and contrast enhancement in medical imaging. In: Fogelman-Soulié F, Gallinari P (eds) ICANN'95. EC2, Paris, pp 41–48
16. Herve T, Dolmazon JM, Demongeot J (1990) Random field and neural information: a new representation for multi-neuronal activity. Proc Natl Acad Sci U S A 87:806–810
17. Demongeot J, Fricot J (1986) Random fields and renewal potentials. Springer Series, NATO ASI Serie F, vol 20, pp 71–84
18. LeCun Y (1986) Learning process in an asymmetric threshold network. Springer Series, NATO ASI Serie F, vol 20, pp 233–240
19. LeCun Y, Bengio Y, Hinton G (2015) Deep learning. Nature 521:436–444
20. Zhao Y, Gao S, Gallinari P, Guo J (2015) Knowledge base completion by learning pairwise-interaction differentiated embeddings. Data Min Knowl Discov 29:1486–1504
21. Dey N, Ashour AS, Bhatt C (2017) Internet of things driven connected healthcare. In: Bhatt C, Dey N, Ashour A (eds) Internet of things and big data technologies for next generation healthcare. Studies in big data, vol 23. Springer, Cham, pp 13–33
22. Elhayatmy G, Dey N, Ashour AS (2018) Internet of things based wireless body area network in healthcare. In: Dey N, Hassanien A, Bhatt C, Ashour A, Satapathy S (eds) Internet of things and big data analytics toward next-generation intelligence. Studies in big data, vol 30. Springer, Cham, pp 3–20
23. Dey N, Ben Abdessalem Karâa W, Chakraborty S, Banerjee S, Salem MAM, Azar AT (2015) Image mining framework and techniques: a review. Int J Image Min 1:45–65
24. Karâa BA, Dey N (2015) Biomedical image analysis and mining techniques for improved health outcomes. IGI Global, Hershey
25. Dash S, Subudhi B (2016) Handbook of research on computational intelligence applications in bioinformatics. IGI Global, Hershey
26. Lytras MD, Papadopoulou P (2017) Applying big data analytics in bioinformatics and medicine. IGI Global, Hershey
27. GtRNA (2017) http://gtrnadb.ucsc.edu/
28. Demongeot J, Hazgui H (2016) The Poitiers school of mathematical and theoretical biology: Besson-Gavaudan-Schützenberger's conjectures on genetic code and RNA structures. Acta Biotheor 64:403–426
29. Demongeot J (2017) Proto-ribosome: a theoretical approach based on RNA relics BioRxiv preprint. http://dx.doi.org/10.1101/182055
30. Demongeot J (2017) "Proto-ribosome" as new game of life. BioRxiv preprint. http://dx.doi.org/10.1101/182055

31. Ramírez-Gallego S, García S, Mouriño-Talín H, Martínez-Rego D, Bolón-Canedo V, Alonso-Betanzos A, Benítez JM, Herrera F (2016) Data discretization: taxonomy and big data challenge. WIREs Data Min Knowl Discov 6:5–21
32. Boukharrou RR, Ilié JM, Saidouni DE (2017) Time and space reasoning for ambient systems. Int J Ambient Comput Intell 8:38–57
33. Mhetre NA, Deshpande AV, Mahalle PN (2016) Trust management model based on fuzzy approach for ubiquitous computing. Int J Ambient Comput Intell 7:33–46
34. Dey N, Ashour AS (2016) Classification and clustering in biomedical signal processing. IGI Global, Hershey
35. Acharjya D, Anitha A (2017) A comparative study of statistical and rough computing models in predictive data analysis. Int J Ambient Comput Intell 8:32–51
36. Sarwar KM, Dey N, Ashour AS (2017) Large scale medical data mining for accurate diagnosis: a blueprint. In: Khan S, Zomaya A, Abbas A (eds) Handbook of large-scale distributed computing in smart healthcare. Scalable computing and communications. Springer, Cham, pp 157–176
37. Singh A, Sharma A, Dey N (2015) Web recommendation techniques-status, issues and challenges. J Netw Commun Emerg Technol 5:57–65
38. Santhi V (2017) Recent advances in applied thermal imaging for industrial applications. IGI Global, Hershey
39. Demongeot J, Hamie A, Glaria A, Taramasco C (2013) Dynalets: a new time-frequency representation of periodic biological signals and spectral data. In: IEEE AINA' 13. IEEE Press, Piscataway, pp 1525–1532
40. Demongeot J, Hamie A, Hansen O, Franco C, Sutton B, Cohen EP (2014) Dynalets: a new method of modelling and compressing biological signals. Applications to physiological and molecular signals. CR Biol 337:609–624
41. Demongeot J, Galli Carminati G, Carminati F, Rachdi M (2015) Stochastic monotony signature and biomedical applications. CR Biol 338:777–783
42. Baconnier P, Benchetrit G, Demongeot J, Pham Dinh T (1984) Morphologie des Cycles Respiratoires: tentative de classification. In: Gallis JL (ed) Actes du IIIème Séminaire de Biologie Théorique. Publications University Bordeaux, Bordeaux, pp 87–102 (http://cat.inist.fr/?aModele=afficheN&cpsidt=9132966)
43. Benchetrit G, Shea SA, Pham Dinh T, Bodocco S, Baconnier P, Guz A (1989) Individuality of breathing patterns in adults assessed over time. Respir Physiol 75:199–210
44. http://home.ccr.cancer.gov/ncifdaproteomics/ppatterns.asp
45. Demongeot J, Hamie A, Laksaci A, Rachdi M (2016) Relative-error prediction in nonparametric functional statistics: theory and practice. J Multivar Anal 146:261–268
46. Rachdi M, Laksaci A, Demongeot J, Abdali A, Madani F (2014) Theoretical and practical aspects of the quadratic error in the local linear estimation of the conditional density for functional data. Comput Stat Data Anal 73:53–68
47. Preda C, Saporta G, Lévéder C (2007) PLS classification of functional data. Comput Statistics 22:223–235
48. http://www.who.int/whosis/icd10/; http://www.who.int/classifications/icf/en/
49. Demongeot J, Virone G, Duchêne F, Benchetrit G, Hervé T, Noury N, Rialle V (2002) Multi-sensors acquisition, data fusion, knowledge mining and alarm triggering in health smart homes for elderly people. CR Biol 325:673–682
50. Virone G, Noury N, Demongeot J (2002) A system for automatic measurement of circadian activity deviations in telemedicine. IEEE Trans Biomed Eng 49:1463–1469
51. Rialle V, Duchêne F, Noury N, Bajolle L, Demongeot J (2002) Health 'smart' home: information technology for patients at home. Telemed J E-Health 8:395–409
52. Le Bellego G, Noury N, Virone G, Mousseau M, Demongeot J (2006) A model for the measurement of patient activity in a hospital suite. IEEE Trans ITB 10:92–99
53. Vuillerme N, Chenu O, Pinsault N, Fleury A, Payan Y, Demongeot J (2008) Can a plantar pressure-based tongue-placed electrotactile biofeedback improve postural control under altered vestibular and neck proprioceptive conditions? Neuroscience 155:291–296

54. Vuillerme N, Pinsault N, Fleury A, Chenu O, Demongeot J, Payan Y, Pavan P (2008) Effectiveness of an electro-tactile vestibular substitution system in improving upright postural control in unilateral vestibular-defective patients. Gait Posture 28:711–715
55. Vuillerme N, Pinsault N, Chenu O, Fleury A, Payan Y, Demongeot J (2009) A wireless embedded tongue tactile biofeedback system for balance control. Pervasive Mobile Comput 5:268–275
56. Chenu O, Payan Y, Hlavackova P, Demongeot J, Cannard F, Diot B, Vuillerme N (2012) Pressure sores prevention for paraplegic people: effects of visual, auditive and tactile supplementations on overpressures distribution in seated posture. Appl Bionic Biomech 9:61–67
57. Fouquet Y, Franco C, Diot B, Demongeot J, Vuillerme N (2013) Estimation of task persistence parameter from pervasive medical systems with censored data. IEEE Trans Mobile Comput 12:633–646
58. Guttmann A, Ouchchane L, Li X, Perthus I, Gaudart J, Demongeot J, Boire JY (2013) Performance map of a cluster detection test using the extended power. Int J Health Geogr 12:47
59. Franco C, Fleury A, Guméry PY, Diot B, Demongeot J, Vuillerme N (2013) iBalance-ABF: a smartphone-based audio-biofeedback balance system. IEEE Trans Biomed Eng 60:211–215
60. Diot B, Hlavackova P, Demongeot J, Vuillerme N (2014) Sensory substitution for balance control using a vestibular-to-tactile device. Multisens Res 27:313–336
61. Rachdi M, Laksaci A, Demongeot J, Abdali A (2014) Theoretical and practical aspects on the quadratic error in the local linear estimation of the conditional density for functional data. Comput Stat Data Anal 73:53–68
62. Demongeot J, Hansen O, Hamie A, Hazgui H, Virone G, Vuillerme N (2015) Actimetry@home: actimetric tele-surveillance and tailored to the signal data compression. In: ICOST'14, Lecture notes in computer science, vol 8456, pp 59–70
63. Demongeot J, Hamie A, Laksaci A, Rachdi M (2016) Relative-error prediction in nonparametric functional statistics: theory and practice. Journal of Multivariate Analysis 146:261–268
64. Demongeot J, Laksaci A, Naceri A, Rachdi M (2017) Local linear regression modelization when all variables are curves. Stat Probab Lett 121:37–44
65. Chan M, Campo E, Estève D, Fourniols JY (2009) Smart homes—current features and future perspectives. Maturitas 64:90–97
66. Demiris G, Hensel BK (2008) Technologies for an aging society: a systematic review of "smart home" applications. Yearb Med Inform 3:33–40
67. Shults MC, Rhodes RK, Updike SJ, Gilligan BJ, Reining WN (1994) A telemetry instrumentation system for monitoring multiple subcutaneously implanted glucose sensors. IEEE Trans Biomed Eng 41(10):937–942
68. Uchiyama A (1995) Endo-radioprobe needs micro machine technology. In: Proceedings of the 6th international symposium on micro machine and human science. IEEE Press, Piscataway, pp 31–37
69. Al-Taee MA, Al-Nuaimy W, Al-Ataby A, Muhsin ZJ, Abood SN (2015) Mobile health platform for diabetes management based on the Internet-of-Things. In: Applied electrical engineering and computing technologies (AEECT'15). IEEE Press, Piscataway, pp 1–5
70. Demongeot J, Noury N, Vuillerme N (2008) Data fusion for analysis of persistence in pervasive actimetry of elderly people at home. In: IEEE ARES-CISIS' 08. IEEE Press, Piscataway, pp 589–594
71. Dittmar A, Axisa F, Delhomme G, Gehin C (2004) New concepts and technologies in home care and ambulatory monitoring. Stud Health Technol Inform 108:9–35
72. http://www.legrand.fr/professionnels/les-appliques-ecoconfort-pour-etablissements medicalises_2795.html

73. Demongeot J, Elena A, Jelassi M, Ben Miled S, Bellamine Ben Saoud N, Taramasco C (2016) Smart homes: data fusion for preventive and therapy education at home; example of type II diabetes. Information 7:50
74. Couturier P, Franco A, Piquart JF, Mansotte J, Montani C, Suarez C, Mollier A, Gucher C, Frossard M, Nicolas L, Jasso Mosqueda G, Mouchet MC, Argentier AM, Bosson JL, Carpentier P, Demongeot J (1996) "Telegerontology": from teleassistance to teleconsultation of elderly people. Mythe or reality? Rev Gériatrie 21:23–31
75. Moreau-Gaudry A, Robineau F, André PF, Prince A, Pauget P, Demongeot J, Payan Y (2006) Utilisation de la substitution sensorielle par électrostimulation linguale pour la prévention des escarres chez les paraplégiques. Etude préliminaire. L'escarre 30:24–37
76. Vazquez-Buenos Aires J, Payan Y, Demongeot J (2003) Electrostimulation of the tongue as a passive surgical guiding system. In: de Almeida AT, Nunes U (eds) ICAR'03. IEEE Press, Piscataway, pp 638–643
77. Tassicker GE (1956) Preliminary report on a retinal stimulator. Br. J. Physiol. Opt. 12:102–105
78. Starkiewicz W, Kuliszewski W (1963) The 80-channel elektroftalm. In: Proceedings of the International congress technology blindness. The American Foundation Blindness, New York
79. Konarzewski T (1992) 100-lecie elektroftalmu. Arch Hist Filoz Med 55:95–99
80. Visell Y (2009) Tactile sensory substitution: models for enaction in HCI. Interact Comput 21:38–53
81. Szkaradek M, Futyma-Ziaja M, Lewicka-Chomont A, Rejdak R (2014) A technology that enables to restore visual function. e-Polish J Vet Ophtalmol. Okulistyka 3
82. Fournier d'Albe EEF (1914) On a type-reading optophone. Proc Roy Soc A Math Phys Eng Sci 90(619):373–375
83. Samsó Dies JM, Vila Bado J, Torres de Gasso JM (1962) Acoustic sensory prosthesis for tactile stimulation. Animal experimentation. Plan and possibilities of the same. Med Clin 39:363–373
84. Samsó Dies JM, Vila Bado J, Torres de Gasso JM (1965) Possibilités d'une prothèse sensorielle acoustique par stimulation tactile cybernetics of the nervous system. In: Wiener N, Schadé N (eds) Progress in brain research, vol 17. Elsevier, Amsterdam, pp 169–178
85. Bach-y-Rita P, Collins CC, Saunders FA, White B, Scadden L (1969) Vision substitution by tactile image projection. Nature 221:963–964
86. http://www.texisense.com/
87. http://www.feetme.fr/
88. Aud MA, Abbott CC, Tyrer HW, Neelgund RV, Shriniwar UG, Mohammed A, Devarakonda KK (2010) Smart carpet: developing a sensor system to detect falls and summon assistance. J Gerontol Nurs 36:8–12
89. Tanaka O, Ryu T, Hayashida A, Moshnyaga VG, Hashimoto K (2014) A smart carpet design for monitoring people with dementia. Prog Syst Eng 330:653–659
90. Andries M, Charpillet F, Simonin O (2015) High resolution pressure sensing using sub-pixel shifts on low resolution load-sensing tiles. In: ICRA'15. IEEE Press, Piscataway, pp 3890–3895
91. Demongeot J, Jelassi M, Hazgui H, Ben Miled S, Bellamine Ben Saoud N, Taramasco C (2018) Biological networks entropies: examples in neural memory networks, genetic regulation networks and social epidemic networks. Entropy 20:36
92. Bourdessol H, Pin S (2008) Practice guide prevention of falls in the elderly living at home. INPES, Paris
93. Guettari T (2014) Détection de présence humaine et évaluation de la qualité du sommeil en établissement d'hébergement pour personnes âgées dépendantes (EHPAD). Télécom Sud Paris Ph.D. thesis, Paris
94. Cavalcante Aguilar PA, Boudy J, Istrate D, Dorizzi B, Moura Mota IC (2014) A dynamic evidential network for fall detection. IEEE J Biomed Health Inform 18:1103–1113

95. http://www.visuresp.com/
96. http://www.login.cl/noticias/1570-ssvq-y-universidad-de-valparaiso-impulsan-innovador-proyecto-de-tele-rehabilitacion-para-pacientes-cardiovasculares
97. Levitzky MG (2008) Using the pathophysiology of obstructive sleep apnea to teach cardiopulmonary integration. Adv Physiol Educ 32:196–202
98. Demongeot J, Demetrius L (2015) Complexity and stability in biological systems. Int J Bifurcat Chaos 25:40013
99. Demongeot J, Bezy-Wendling J, Mattes J, Haigron P, Glade N, Coatrieux JL (2003) Multiscale modeling and imaging: the challenges of biocomplexity. In: Proceedings of the IEEE society, vol 91, pp 1723–1737
100. Jelassi M (2017) Modélisation, simulation et analyse multiéchelle de réseaux sociaux complexes: Application à l'aide à la prévention des maladies contagieuses. Ph.D. Thesis. University Grenoble Alpes, Grenoble
101. Mall R, Rawi R, Ullah E, Kunji K, Khadir A, Tiss A, Abubaker J, Dehbi M, Bensmail H (2017) Application of high-dimensional statistics and network based visualization techniques on arab diabetes and obesity data. BioRxiv preprint. http://dx.doi.org/10.1101/151621
102. Demongeot J, Jelassi M, Taramasco C (2017) From susceptibility to frailty in social networks: the case of obesity. Math Pop Stud 24:219–245

Chapter 2
Aerial and Satellite Imagery and Big Data: Blending Old Technologies with New Trends

J. Salazar Loor and P. Fdez-Arroyabe

Abstract Over the past decades, the successful employment of aerial and satellite imagery and remote sensing (RS) data has been very diverse and important in many scientific fields. Firstly, a brief review of RS history is presented in section one. Then, basic properties, which are also challenges, of RS big data are concisely discussed. Volume, variety and velocity are mainly described as characteristics of RS big data while variety, value and visualization are primarily denoted as new challenges. The third section is concentrated on justifying the relevance of RS big data in today's society and the needs to integrate it with other kind of data sources to develop useful services. In this sense, a special section is dedicated to Copernicus initiative and some case studies of specific applications are also shown. Finally, some general conclusions are presented paying attention to the spatial nature of RS big data, which gives it a special added value in the new digital era.

Keywords Aerial and satellite imagery · Remote sensing · Spatial big data Integration

2.1 History of Remote Sensing

For us to know the exact moment when mankind first used remote sensing (RS), we would have to go back to France in the 1830s, when Louis Daguerre invented the daguerreotype process of photography. Initially, photography was simply an artistic instrument that helped reproduce the nature, landscape or people's lives during

---

J. Salazar Loor
Natural and Environmental Sciences Department, International University SEK, UISEK, Calle Alberto Einstein s/n y 5ta. Transversal, 170302 Quito, Ecuador
e-mail: jose.salazar@uisek.edu.ec

P. Fdez-Arroyabe (✉)
GeoBioMet Research Group, Geography and Planning Department, University of Cantabria, ETSI, Avda. Castros s/n, 39005 Santander, Cantabria, Spain
e-mail: fernandhp@unican.es

N. Dey et al. (eds.), *Big Data for Remote Sensing: Visualization, Analysis and Interpretation*, https://doi.org/10.1007/978-3-319-89923-7_2

Fig. 2.1 Honoré Daumier (French, 1808–1879). *Nadar Élevant la Photographie à la Hauteur de l'Art*, May 25, 1862. Lithograph on newsprint, Sheet: 13 1/4 x 9 3/4 in. (33.7 x 24.8 cm). Brooklyn Museum, Frank L. Babbott Fund, 51.4.3

those years [1]. However, the uses of photography diversified and Gaspard-Félix Tounachon, best known for his pseudonym Nadar, took the world's first aerial photograph in 1858 from a tethered balloon (Fig. 2.1) [2]. Even though his aerial photos have since been lost, Nadar's innovation is considered the first attempt to study Earth's surface using RS and has inspired many subsequent works [3].

The introduction of photographic cameras in powered aircrafts was the next logical step in reaping the benefits of remote sensing. Hence, Wilbur Wright took several motion pictures of a location near Centocelli [4]. These pictures were considered the first aerial photos taken from an airplane. The maneuverability of the airplane, among other characteristics, attracted the interest of institutions, governments and developers, which saw them as advantages for the systematic use of aerial cameras [5]. In 1910, the International Society of Photogrammetry and Remote Sensing (ISPRS) was founded with the purpose of bringing together professionals and specialists of photogrammetry and remote sensing from all around the world [6]. During the 1920s, aerial photography was almost restricted to military applications [3]. The use of aerial photography rapidly matured during World War I, as reconnaissance aircraft were equipped with cameras to acquire routine photographs [7]. Data obtained from these images were really valuable for the parties involved in the conflict [3].

During the Inter-War years (1919–1938), aerial photography experienced many improvements. Photogrammetry, for example, became a new scientific field focused on creating techniques that allowed to make accurate measurements from aerial photographs [5]. New photogrammetry instruments were created as well, while photogrammetry products were mainly employed by government programs for topographic mapping, soil survey, geologic mapping, forest studies and agricultural statistics [3]. Several innovators contributed to the development of new photogrammetry techniques and instruments as well. Among them were Sherman Fairchild, Talbert Abrams, and Willis Thomas Lee. A successful aerial camera designer and businessman, Fairchild founded numerous companies along the way, while Abrams also introduced new innovations in camera design. Thomas Lee, for his part, wrote "*The Face of the Earth as Seen from the Air*" in 1922, where he described possible applications of aerial photography in a wide range of disciplines [8].

The Great Depression (1929–1939) generated economic, financial and environmental crisis in many countries, but it was also a window of opportunity for photogrammetry and aerial photograph to become an important tool of governmental institutions [9]. For instance, the U.S. Department of Agriculture (USDA) made efforts to use aerial photography for economic planning and development [10]. Others governmental uses included studies about soil erosion and reliability of water supplies [3].

Remote sensing took a new step during World War II [11]. The wavelength of the electromagnetic spectrum expanded to include infrared and microwave radiation (Fig. 2.2). Previously, the spectrum only showed visible radiation, which is the portion that is visible to the human eye. Nowadays, researchers provide theoretical and practical methods for using the nonvisible spectrum in further applications [11].

Most countries had already noticed the success of strategic photointerpretation by the end of World War II. Remote sensing was indeed an effective way to acquire reliable information from other territories [3]. Consequently, the United States government developed a series of surveillance programs and instruments. In 1955, the Lockheed U-2 reconnaissance aircraft became the first of such advances. The Corona strategic reconnaissance satellites came shortly after. They operated

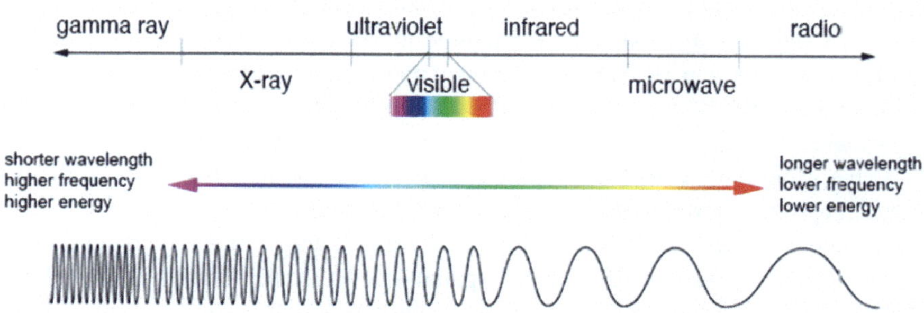

Fig. 2.2 Comparison of wavelength, frequency and energy of the electromagnetic spectrum. *Source* Imagine the Universe!—NASA

between 1959 and 1972, providing photographic surveillance from space [12]. These inventions allowed the US to detect the installation of nuclear-armed Soviet missiles on Cuba in 1962 [3].

Governmental institutions were not alone in RS applications and advances. There were other civil organizations, too. Reserved for surveillance purposes until then, the launch of the TIROS-1—the world's first successful low-Earth orbital weather satellite—and the release of old military instruments for civil uses broadened the spectrum of RS applications [13]. Robert Cowell's work is one worthy of mention. In 1956, he applied color infrared film—a technique that helped detect camouflage in World War II—to identify problems and diseases in small grain cereal crops, improving studies in agricultural science and bringing new dimensions to remote sensing [3].

Although the non-visible spectrum had been studied and used since the 1920s, the term "remote sensing" wasn't coined until 1950 [3]. Evelyn L. Pruitt, a geographer for the U.S. Department of the Navy (DoN), realized that the term aerial photography didn't describe other techniques and imagery which work outside the visible spectrum. Pruitt said that remote sensing is "the science—and art—of identifying, observing, and measuring an object without coming into direct contact with it" [14]. This fresh approach to remote sensing brought new opportunities to explore its applications. Consequently, the National Aeronautics and Space Administration (NASA) created a remote sensing research program during the 1960s. The National Academy of Sciences (NAS), for its part, examined the use of remote sensing in agriculture and forest studies [3].

The 1970s marked a new milestone for remote sensing. In 1972, NASA launched Landsat 1, the first satellite of the United States Landsat program. Even though its contribution was not fully recognized at the time, Landsat 1 paved the way for many advances in remote sensing. Landsat 1 had the ability of routinely taking multispectral satellite imagery of Earth [15]. This fact sparked interest in multispectral data from scientists and researchers alike. Additionally, the Landsat program was the catalyst behind the quick and broad expansion of RS digital image analysis. Up until then, analysis of RS data, including images and aerial photographs, was mostly visual [16]. A standardized digital format as well as the digital availability of Landsat imagery, distributed in transparency or paper print format, allowed development of techniques, tools and computer solutions for multispectral imagery analysis [17]. As a result, Landsat finally became an example to other Earth observation satellites. Landsat 2, launched in 1975, had better resolution, while commercial satellites such as GeoEye or IKONOS offered new RS platforms. Scientists also improved techniques and computer programs for multispectral data analysis [18].

At the same time, early Geographic Information Systems (GIS) evolved thanks to the integration of different sources such as relational and hierarchical Data Base Management Systems (DBMS), graphic design systems like Computer Aided Design (CAD) as well as RS-oriented spatial analysis capabilities (Fig. 2.3) [19]. Thanks to its GIS color-graphic display capabilities, electrostatics and inkjet plotters became popular printing devices too. Binding data base and interactive graphics systems was another milestone for the development of GIS algorithms and

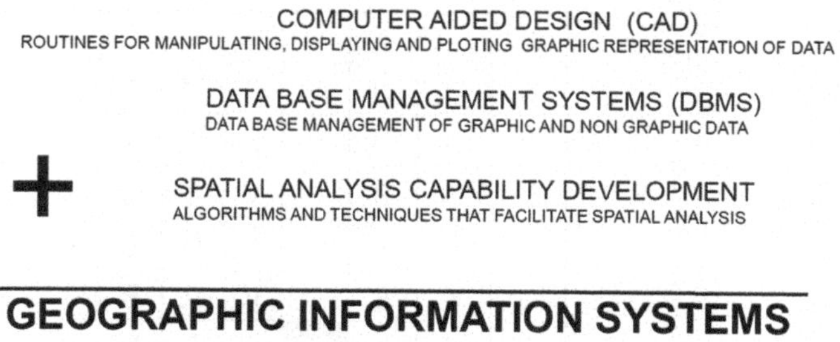

Fig. 2.3 Geographic information systems (GIS) integrate data from a number of different sources

techniques to visualize and analyze RS data sets. NASA and the Landsat program promoted GIS development, popularizing the use of Landsat data and helping the emergence of commercial software solutions like ERDAS or Arc-Info [20].

Throughout the 1980s, remote sensing broke new ground. NASA's Jet Propulsion Laboratory (JPL) developed AVIRIS, an imaging spectrometer for detailed Earth remote sensing [3, 21]. In 1999, NASA launched the Terra (EOS AM-1) satellite. It was the first multi-national scientific research satellite designed to monitor the state of Earth's environment [22]. The appearance of geospatial data, which primarily involves RS technologies, and Global Positioning Systems (GPS) were also relevant breakthroughs [3]. Geospatial data is any kind of information that has geographic components such as coordinates, addresses, locations, ZIP codes, etc. GPS systems are able to obtain images with high-accuracy positioning and integrate them with other spatial information [23].

Public services offering this type of information were another pivotal moment in remote sensing history. At the beginning of the 21st century, The Internet increased its influence and companies began to develop services that provided RS imagery to the public all around the world [24]. In 2005, Google launched Google Earth, a computer program that provides imagery to a public more familiar with the Internet and the World Wide Web (www) than with RS techniques and theories [3, 24].

2.2 Remote Sensing Big Data Properties and Challenges

Basically, RS data is "the use of space-borne and airborne sensor technologies to detect and classify objects on Earth, as well as some other data acquisition measurements" [25]. RS data is currently an important source of information for the study and learning of the world. Several of its characteristics are expressed through a host of key concepts (Fig. 2.4) that allow us to consider RS data as big data. The number of concepts varies depending on the authors. Volume, velocity and variety are always mentioned as the main dimensions or key properties of big data [26]. However, veracity, visualization and value, as new challenges, give a much more complex theoretical framework to big data.

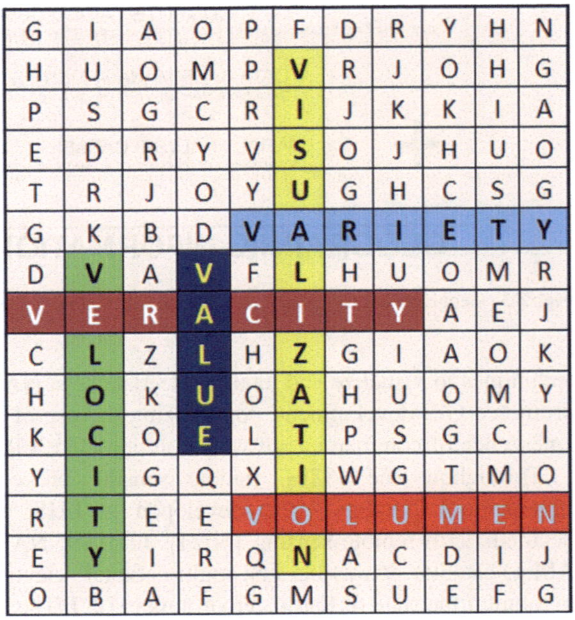

Fig. 2.4 Properties and challenges of RS big data

The six mentioned versus can be considered properties and challenges of RS big data simultaneously. They refer to a continuous process of capturing, storing, managing, analyzing and visualizing RS big data with multiple scientific and commercial purposes.

2.2.1 "V" for Volume

In first place, RS data possesses a vast amount of useful information that's open to the public [25]. Secondly, the satellite observation networks displayed around the world have more than 200 on-orbit satellite sensors [25, 27]. Nowadays, NASA's Earth Observing System Data and Information System (EOSDIS) has stored a 7.5 petabytes remote sensing imagery archive that delivered about 4.5 million gigabytes of data by 2012 [27, 28].

This huge amount of information presents both an advantage as well as problem for scientists. Even though they have more material to work with, storage requirements for all this information are huge and a challenge too difficult to accomplish for many computers [25].

NASA is probably the biggest distributor of RS data in the world, collecting 1.73 gigabytes of such data every second [29]. Concerning data management, this number implies that storage centers constantly need capacity upgrades [25]. Data centers need rigorous requirements on storage, processing and transmission capacity to properly manage large volumes of RS data. Other essential features

Fig. 2.5 Data measurement chart

1 Byte	8 bits
1 kilobyte (KB)	1024 B
1 Megabyte (MB)	1024 KB
1 Gygabyte (GB)	1024 MB
1 Terabytee (TB)	1024 GB
1 Petabyte (PB)	1024 TB
1 Exabyte (EB)	1024 PB
1 Zettabyte (ZB)	1024 EB
1 Yottabyte (YB)	1024 ZB
1 Xentabyte (XB)	1024 YB
1 Wektabyte (WB)	1024 XB

include a proper number of nodes, high-speed internal network, effective cooling systems, and a good backup storage capacity [28]. Only data centers that satisfy high energy efficiency, stability, safety, capacity and redundancy can optimally ensure these operations [25].

On the other hand, small remote sensing operators have their own issues in getting RS data ready for analysis [25]. Numerous pre-processing techniques are done to ensure having suitable data. These include radiometric, geometric and inconsistency correction, as well as image [27]. Additionally, machines designed for managing RS data are expensive. A 128 GB RAM server, which handles 100 GB and processes a program in less than one minute, cost over $15.000 [30]. It seems like a lot of information, but this machine can't deal with NASA or European Space Agency (ESA) data [31, 32].

As RS data can manage big volumes of information, it clearly falls into big data category. Figure 2.5 shows units used to measure data volume, which are constantly evolving thanks to the continuous increase of data production.

2.2.2 *"V" for Variety*

Variety refers to the number of types of data. However, it's important to differentiate "*remote sensors data*" from "*remote sensing data*" before elaborating on variety in relation to the latter. Remote sensors (i.e. devices connected to the Internet like mobile phones, tablets, traffic cameras, or monitoring instruments) collect zettabytes of unstructured and semi-structured information from multiple users. Conversely, remote sensing data is specifically linked to mainly structured information such as images obtained by satellites sensors (Fig. 2.6).

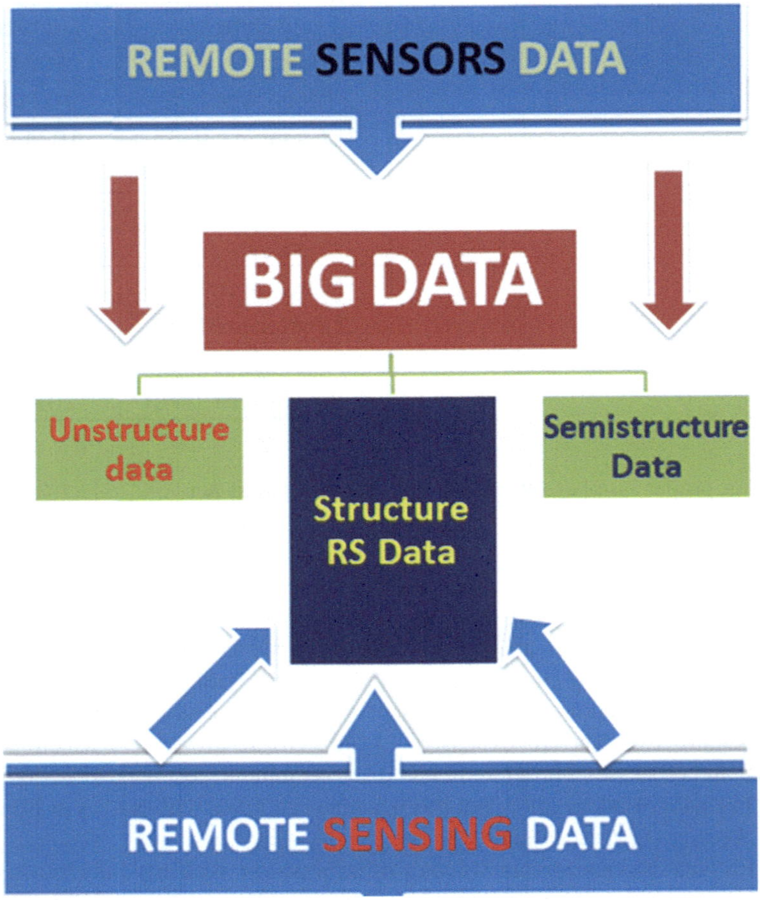

Fig. 2.6 *Remote Sensors* data versus *remote sensing* data

Understandably, the concept of variety takes a different connotation when connected to *remote sensing* data. To some extent, variety depends on the different resolution (spectral, temporal, spatial, and radiometric) of the captured data.

Under this approach, RS data variety is also enormous. There are approximately 200 satellite sensors with an oversize variety of spatial, temporal, radiometric and spectral resolutions [25]. For instance, the Earth Observing System Data and Information System (EOSDIS) has sensors with very diverse temporal and spectral resolutions. They are transported in Far-Space satellites, Near-Space satellites and airborne platforms [25].

Usually, remote sensing sensors are created to study a specific topic or issue. There are sensors that capture information for environmental monitoring, land processes, atmospheric factors, hydrology, oceanography and other fields. Since these instruments have different purposes, they also have different characteristics.

Therefore, managing and integrating all the information isn't an easy task, even if it's mainly structured data [3].

Between active and passive devices, there is a host of remote sensors in the market, including gamma ray spectrometers, multispectral scanners, imaging spectrometers, satellite line cameras, aerial cameras, video cameras, thermal scanners, passive microwave radiometers, laser scanners, radar altimeters, and imaging radars [30].

Raw data is stored in data sets. NASA archives have 7000 types of data sets structured in different standard formats such as HDF, netCDF, GeoTIFF, FAST, ASCII and many more. Usually, these formats have different structural metadata [25].

The spatial resolution of some of the most famous satellite sensors ranges between 0,4 m and 1 km: HJ-1C (5 m), SPOT-5 (2.5 m), IKONOS (1 m), QuickBird (0.61 m), Orbview-5 (0.41 m), LandSat-5 TM (30 m), LandSat-7 ETM + (15 m), LandSat-8 OLI (15 m), NOAA (1 km), and MODIS (250 m) [25]. Temporal resolution is also wide and fluctuates between a few hours, months or years. The temporal frequency of the MODIS LST system can be daily, weekly or monthly [32]. Finally, spectral resolution is quite broad as well. The remote sensing Hyperion sensor consists of 220 spectral bands with 10 mm resolution; Landsat sensors (OLI and TIRS) combine bands that go from ultra-blue to far infrared; and the QuickBird Satellite Sensor works with four bands ranging from blue to near infrared [25].

2.2.3 *"V" for Velocity*

Velocity refers to the frequency of incoming data and the speed at which is generated, processed and transmitted. However, these data pre-processing steps require users, especially for those living in developing countries, to have expensive software, technological skills and hardware resources.

Facebook's data warehouse stored upwards of 300 PB of Hive data by 2014, with an incoming daily rate of about 600 TB. Danish researchers at the Technical University of Denmark used a new type of optical fiber and data transfer speed reached a maximum world value of 43 Tbsps. New technical findings allow transfer velocity to keep constantly increasing. And since speed and cost are directly related, it has sparked a fierce race between companies like AWS, Windows and Google for the most competitive data transfer prices in the market.

Data processing velocity and rates have become also a major issue in big data storage and analytics. Currently, there's an increasing pressure on analytics teams to find faster ways to solve much more complex multi-structured data environments in a shorter time (Fig. 2.7). Data velocity is not only related to machines and electronic devices but to human understanding and transforming data into useful applications as well [33].

The ability to make real-time decisions on complex issues is one of the main properties and biggest challenges for RS big data because they often come with risks

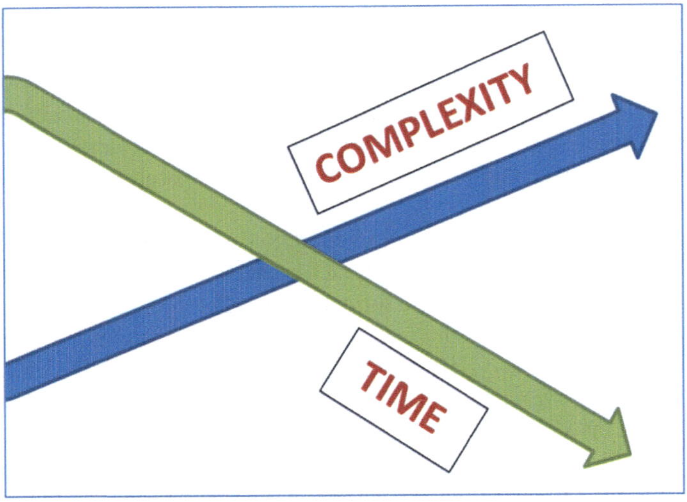

Fig. 2.7 Time versus complexity in big data analytics

such as big economic losses, epidemiological risks, or life-threatening situations. As a result, expert and decision-making systems—both based on huge amounts of heterogeneous data where velocity is a key parameter—are under a lot of pressure. Complexity is based on interdisciplinary and transdisciplinary analysis and processing time can be essential dealing with complexity to make on real time right solutions.

2.2.4 *"V" for Veracity*

The veracity of RS data needs proof testing before being used for specific purposes. This is one of the principal challenges of any big data including RS one. Verifying how data was collected, cleaned, transformed and integrated are essential steps to validate its quality. Figure 2.8 shows the basic elements of data veracity.

Data veracity depends on data reliability, data accuracy and data governance. Initially, RS data is high in reliability due to its structured information and associated uncertainties linked to sensor calibration and resolution, whether spatial, temporal or radiometric. These are usually defined by the institutions and enterprises in charge of collecting, cleaning and transforming data. The lack of unstructured data in RS data increases data reliability [33, 34].

Data accuracy is also essential to determine the quality of RS data. If they are not detected in the early stages, voids or errors found in the data sets as well as the lack of accuracy in spatial and non-spatial components of the RS data can become uncertainties which then make their way into the analytical process. It has been proven that data uncertainties increase when data sets collected by scientific institutions are transferred to the Internet of Things (IoT) [26].

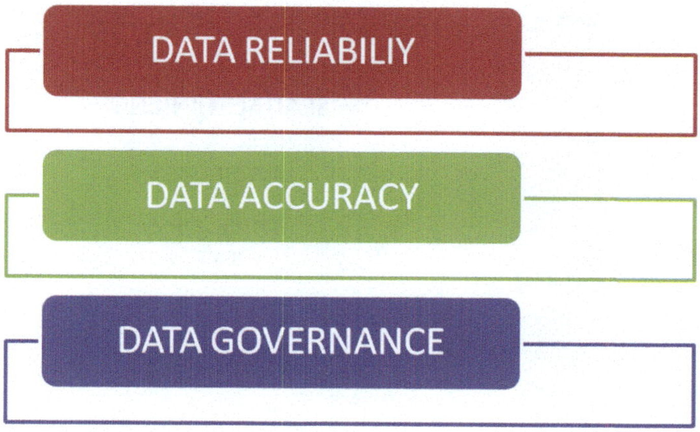

Fig. 2.8 Basic challenges of data veracity

Finally, data governance provides a much-needed balance between data democratization (Open Source) and information privacy (e.g. medical records, bank accounts, private conversations, etc.). Strategic RS data sets are the key to guarantee personal and global security. Cybercrimes are known to prosper without a proper data governance system to prevent them.

RS data veracity is targeted-oriented, which means that several pre-processing techniques (e.g. radiometric, geometric and inconsistency correction, image enhancement, etc.) applied to raw data must be different for each specific research project or analytical process [35].

2.2.5 "V" for Value

Also known as geospatial data or geographic information, spatial data is characterized by two components: *spatial dimension* (i.e. measurement-based geolocation using a geographic coordinate system) and *attribute* (i.e. numeric value that represents the radiometric dimension of each specific location). Spatial data provides an added value to RS data and big data analytics (e.g. spatial patterns, correlations of social and natural phenomena with geographical features) [36].

Indeed, information systems incorporate spatial information systems that use geographic information (Fig. 2.9). Therefore, every satellite image contains spatial data of different natural and artificial ecosystems (e.g. oceans, forests, cities, zoos) and RS techniques are applied to each one of them.

Furthermore, spatial thinking has become a new challenge in the quest of big data analytics to understand the true nature of reality, and Earth observation (EO) programs such as the Copernicus Initiative are now based on capturing spatiotemporal data.

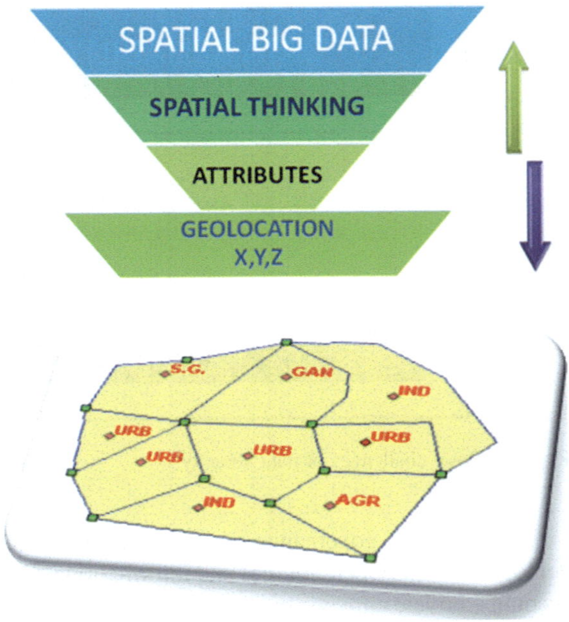

Fig. 2.9 The added value of spatial data

Nevertheless, there are still issues left to dealt with regarding the spatial components of remote sensing in big data systems. According to Vatsavay and Chandola, "*spatial components need to evolve in order to meet the challenges posed by big spatial data and spatiotemporal data*" [36]. This statement takes more relevance when raw RS data is transformed into vector data (i.e. points, lines and polygons) with geometric and topologic properties. In such cases, spatial database principles must be incorporated to extract the added value of these properties through spatial big data analytics.

2.2.6 "V" for Visualization

Visualization is yet another challenge for big data [25]. There are five main visualization problems concerning the analysis of diverse and heterogeneous RS data [35]:

- Noise in visualization: most of the objects in data sets are too relative to each other that observers can't distinguish between them.
- Large image perception: human perception loses its ability to acquire any valuable information from data visualization.
- Loss of information: decrease visibility in data sets can generate diminishing returns in information.

- High performance requirements: dynamic and high-performance visualization require a continuous increase of computing resources.
- High rate of image change: observers can't react to the high number of data changes or its intensity display.

In consequence, the development of accessible, user-friendly tools that help visualize large volumes of data and acquire helpful information is a major objective going forward. Scientists from the "Virtual Laboratory of Remote Sensing Time Series" have already developed "virtual-globe based geo-browsing tools that provide free access to satellite-derived high spatial resolution images and geographical" [37]. Using MODIS data sets, these scientists have agreed to create validated MODIS products with high radiometric and geometric quality, as well as developing a new processing and visualizing standard of MODIS RS data [35, 37]. But like any other project, it has had to face several challenges, including the creation of filtering procedures like the wavelet transform, huge efforts in big-data computing (60 days of processing time to construct a data set composed of 500 million EVI2 time series), and the integration of processed MODIS data sets to virtual globe programs such as Google Maps, Google Earth and Microsoft Virtual Earth to make up for the lack of optimal computer architecture for data storage [37].

2.3 Importance, Needs and Applications of RS Big Data

The importance of digital era and RS big data can be compared to other great episodes in human beings history such as the industrial revolution. One might say that remote sensing is to the information society what Henry Ford was to the automotive industry. Never before had so many cars had been produced in such a short period of time. Then again, never before had digital data been so abundant in such a short period of time. Accordingly, big data should be seen as a revolution based on the ideas of "*mass production of data*" and "*consumption of massive data-related services*".

Understanding this new era is essential in order to define appropriate needs in our societies at all levels from education systems to commercial activities, food delivery, water management or health care among many other sectors. Nowadays, remote sensing has become the main source to acquire data of Earth. Big data industry can be understood as a manufacturing factory which aims to achieve higher productivity as well as deliver multiple services to users by standardizing the capture, storage, management, visualization, analysis and transformation of spatial data. To do this, we could say, it uses assembly lines, breaks work down into smaller, more manageable tasks, and must combine artificial intelligence and machine learning with innovative thinking and versatility and efficiency of its workers.

Clear goals are needed before data is collected. Otherwise, the utility of data sets can be easily questionable. The size, structure and consistency of data must be known in order to facilitate its management and the use of ETL and OLAP tools. Additionally, a plan to deal with data inconsistencies and data verification must be designed in advance. Finally, streaming data demands new technical skills for the development of real-time analysis [38, 39].

The need for data warehouses to incorporate books, documents, videos, images, and other sources of semi or unstructured data with relevant information to business and science [40] has made it imperative to create new big data technologies. Currently, the evolution of computing technology allows users to manage large volumes of data without supercomputers using tools like NoSQL, Google BigTable, DataStream Management Systems (DSMS), Simple DB, Memcache DB, or Voldermort [41].

However, big data management and analytics rely on Doug Cutting's Hadoop, which is a collection of Open-Source projects with multiple components [42], including HCatalog, HDFS, Hive, Zookeeper, Pig, Mahout, Oozie, Avro, Chukwa, and Flume.

2.3.1 Copernicus Initiative

Today, users are demanding updated and easily accessible geospatial data that provides useful and helpful information [43] such as monitoring systems that deliver solutions to current global environmental problems (e.g. climate change, desertification, biodiversity loss, etc.). The European Space Agency (ESA) has heard these demands and has been developing the Global Earth Observation System of Systems (GEOSS), better known as the Copernicus Project.

The Copernicus initiative has as main purpose "*to provide accurate, timely and easily accessible information to improve the management of the environment, understand and mitigate the effects of climate change, and ensure civil security*" [43]. Coordinated and managed by the European Commission (EC) and implemented in partnership with all member States, the Copernicus programme is probably the biggest initiative in the world to make RS data freely and universally available to users.

Copernicus initiative is served by a group of satellites—known as the Sentinel families—and missions in order to generate data and services. Information in situ is also collected by Copernicus initiative through ground stations or different sensors on the ground, air or sea. There are many tangible examples where Copernicus initiative benefits can be found such as in the agriculture, forestry and fisheries sectors; biodiversity and environmental protection; climate and energy; civil protection and humanitarian aid; public health; tourism, transport and safety; urban and regional sustainable planning.

Copernicus Services (CS), for its part, addresses six main domains or thematic areas (Atmosphere, Marine, Land, Climate, Emergency and Security). These domains are presented graphically in Fig. 2.10 considering the hidden value of the spatial and geographic component of the information they are collecting.

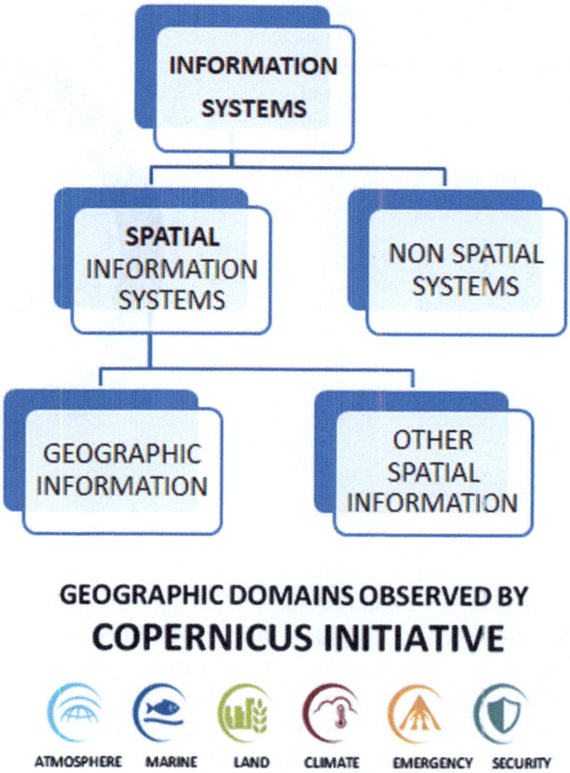

Fig. 2.10 Spatial dimensions of copernicus initiative domains

However, these types of projects need new paradigms that are capable of transforming complex RS imagery into ready-to-use information. And the Copernicus leadership integrated the Object-Based Image Analysis paradigm (OBIA) in response. According to them, OBIA has the "potential to integrate different techniques of processing, retrieving and analysing multi-resolution data from various sensors". It certainly matches OBIA's own purpose, which is to "provide adequate and automated methods for the analysis of very high spatial resolution (VHSR) imagery by describing the imaged reality through the use of spectral, textural, spatial and topological characteristics" [43]. Although OBIA works better than pixel-based approaches, its application is not flawless: current commercial object-based software does not provide simple options to operators, it has problems when processing huge data sets, slight changes in criteria selection could affect segmentation, there is not enough theoretical basis of what OBIA is and its components, and scale and hierarchical relations between objects can be confusing, especially when derived from different resolutions [44].

Object delimitation and evaluation is another of OBIA's problems. Usually, delimitation methods like region-based segmentation algorithms produce areas that share certain conditions of homogeneity (e.g. same spectral respond, compactness, roughness, etc.) [43]. But these methods show their limitations when they have to

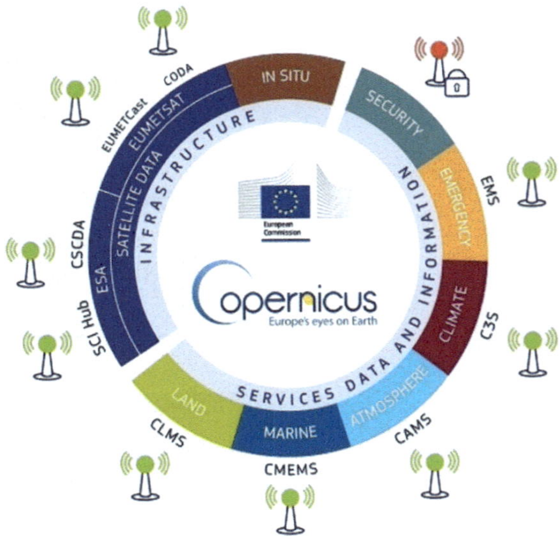

Fig. 2.11 The copernicus space infrastructure. Copernicus a programme of the European Union, managed by the European Commission *Source* http://copernicus.eu/data-access

demarcate high-contrast and regular appearing objects such as an orchard, a mire or a hummock—an easy task for a well-trained human eye [43]. To avoid it, experts suggest the introduction of hybrid techniques that combine computer-based automated delineation of basic units with advance high-level aggregation by a human specialist [43, 44]. The Copernicus Space Infrastructure (CSI) is an example of providing easy access to RS data (Fig. 2.11).

2.3.2 Applications of RS Big Data

The importance of RS data applications is widely known. Studies of biodiversity, nature conservation, food security, deforestation impact, desertification monitoring or urban sprawl are some of the field where RS data employment has had great success. In this aspect, social and environmental necessities such as environmental monitoring requirements, conservation goals, spatial planning enforcement, or ecosystem-oriented natural resources management are driving to develop solutions which produce tangible information from RS data [45].

One example of the advances in the use of RS data is eBird. It is a project which objective is documenting "*bird observations made by recreational and professional bird watchers .... and providing rich data sources for basic information on bird abundance and distribution at a variety of spatial and temporal scales*" [46]. eBird database has more than 9.5 million bird observations and recollects RS data of multiple US institutions.

Between its products, eBird have been developing the Spatio Temporal Exploratory Model (STEM) which relate environmental features (habitat, climate or elevation) and bird watchers data [47]. STEM used remote sensed information from the National Land Cover Database and vegetation phenology from MODIS to build "*a novel methodology for adding essential spatiotemporal structure to existing static species distribution models without requiring detailed specification of the underlying dynamic processes*" [47, 48]. This method restricts static species distribution models to their own spatial and temporal extent and generates over 100 dynamic bird species distribution models with the help of RS data.

Fishery management has been a preoccupation to nations for decades. The United Nations Food and Agriculture Organization (FAO) has made estimations that one-quarter of fish stocks have been overexploited [49]. These reports have created a global concern to improve fishery management and make it more sustainable. In this context, science and technology innovations will play an important role to enhance fisheries management efforts.

A specific aspect where science and technology innovations have a potential is ship tracking using RS data as a surveillance method [50]. For example, Day/Night Band (DNB) imagery of Visible Infrared of Imaging Radiometer Suite (VIIRS) has been used to detect vessel traffic density in regions such as South China Sea or East China Sea [51, 52]. These products are helping to improve other surveillance systems like Ship Automatic Identification System (AIS), which is not mandatory to employ for small vessel or can be stopped for security reason, or Synthetic aperture radar (SAR) imagery, that required a large power supply to operate [51].

Urban sprawl has also been a big concern to scientists and developers. This phenomenon has produced changes in lifestyle in urban inhabitants. Additionally, urban citizens have created new ways to live and recognize the city. From this point of view, urban and peri-urban agriculture has been constituted as an alternative of resources to about 800 million people globally [53].

In developing nations, a primary necessity to successfully implement urban agriculture initiatives are a better planning based on accurate geospatial information [54]. RS data provides spatially consistent data which cover large areas with high spatial detail and high temporal frequency [53]. For instance, RS data helps urban planners in Shanxi, China where have been developing farm monitoring system and establishing an accurate farmland monitoring in over 95 percent of ranches with the help of RS data [55].

2.4 Conclusions

The massive and diverse nature of RS data represents both a great challenge and an enormous opportunity for researchers to created new methods and techniques in many scientific fields. Additionally, the flexibility shown by RS data to work with other kinds of geospatial information is an excellent opportunity for the development of new applications.

On the other hand, researches still have to solve some problems, including new paradigms and advance algorithms, for them to be able to fully integrate RS data with big data. RS data properties such as volume, velocity variety and veracity, diversity of resources, methodologies and its inherent complexity—who some people say represent improvements in the acquisition of new information—are also challenges that researchers and scientists have to overcome in order to seize every aspect of RS data.

For example, research and development services see an enormous potential in RS data variety linked to different resolutions, (spatial, spectral, radiometric and temporal) but it's also very complex to work with the multiple resolutions and a vast diversity of digital formats in which data is generated. Nevertheless, there is no doubt that RS data is the main source of geographic information in the world at present.

Regarding value, its spatial components are the key to obtain added value from RS big data in the future. New spatial analytics must be developed to acquire new results and spatial innovative thinking must be promoted among young researchers.

Visualization is another front that needs a lot of work. Displaying graphically complexity of RS big datasets is an outstanding task to be developed far from the traditional methods of visualization. Tools such as GeoGlobes show great potential despite its problems with data integration. However, further research is necessary to improve its technology.

Veracity of RD big data is a transdisciplinary issue. According to researchers, it validates information and gives added value to their investigations. Accuracy of RS big data is linked to spatial resolution at which data are collected from sensors. Data governance is probably another key aspect of RS big data in the present. Access to information must be legally controlled to avoid citizens privacy can be lost.

The importance of RS big data is clearly justified through initiatives like Copernicus one at the European Union in the context of a second industrial revolution based on digital data and devices. As it happens in any revolution, big and sudden transformations demand quick actions and answers at all levels. It is important to define future needs at the education systems and universities in order to lead these coming changes.

Finally, based on its characteristics, it can truly be stated that RS is indeed part of big data. Any future application based on RS big data will have to be able to work globally to developed customized or personalized services at an individual scale. This integration of multiple spatial scales is, perhaps, the most challenging aim of working with RS big data.

References

1. Davenport A (2000) The history of photography: an overview, 2nd edn. The University of New Mexico Press, Albuquerque
2. Barber M, Wickstead H (2010) One immense black spot': aerial views of London 1784–1918. Lond J 35:236–254
3. Campbell JB, Wynne RH (2011) Introduction to remote sensing, 5th edn. The Guilford Press, New York

4. Butler MJA, Mouchot MC, Barale V, LeBlanc C (1988) The application of remote sensing technology to marine fisheries: an introductory manual. Food and Agriculture Organization of United Nations, Rome
5. Gosh S (1981) History of Photogrammetry. Laval University, Québec
6. Schenk T (2005) Introduction to photogrammetry, 1st edn. The Ohio State University, Columbus
7. Stichelbaut B (2006) The application of First World War aerial photography to archaeology: the Belgian images. Antiquity 80:161–172
8. The Professional Aerial Photographers Association (2017) History of aerial photography
9. Monmonier M (2002) Aerial photography at the agricultural adjustment administration: acreage controls, conservation benefits, and overhead surveillance in the 1930s. Photogramm Eng Remote Sens 68:1257–1262
10. Rango A, Havstad K, Estell R (2011) The utilization of historical data and geospatial technology advances at the Jornada experimental range to support Western America ranching culture. Remote Sens 3:2089–2109
11. Cracknell A, Haynes L (1991) Introduction to remote sensing, 2nd edn. Taylor & Francis Ltd., London
12. Ruffner K (2017) Corona: America's first satellite program. Central Intelligence Agency, Washington, DC
13. NASA Science Website (2016) TIROS: The television infrared observation satellite program. In: NASA Science Website
14. Graham S (1999) Remote sensing: introduction and history. In: NASA Earth Observatory. https://earthobservatory.nasa.gov/Features/RemoteSensing/
15. Mack P (1990) Viewing the earth: The social construction of the landsat satellite sytem. The MIT Press, London
16. NASA Landsat Science (2017) History: from the beginning. In: NASA Landsat Science
17. Van Wie P, Stein M (1976) A landsat digital image rectification system. Greenbelt
18. Patra P (2010) Remote sensing and geographical information system (gis). Assoc Geogr Stud
19. Antenucci JC, Brown K, Croswell PL, Kevany MJ, Archer H (1991) Geographic information systems. A guide to the technology. New York
20. Foresman T (2010) GIS, History of geographic information systems. Encycl Geogr 1281–1284
21. NASA Jet Propulsion Laboratory (2010) AVIRIS—airborne visible/infrared imaging spectrometer—general overview. https://aviris.jpl.nasa.gov/aviris/
22. NASA Terra—The EOS Flagship (2017) Terra Instruments|Terra. https://terra.nasa.gov/about/terra-instruments
23. Mohamed B, Werner K (2007) Geospatial information bottom-up: a matter of trust and semantics. In: Fabrikant SI, Wachowicz M (eds) *The European information society*. Springer, pp 365–387
24. Farman J (2010) Mapping the digital empire: Google earth and the process of postmodern cartography. New Media Soc 12:869–888
25. Ma Y, Wu H, Wang L, Huang B, Ranjan R, Zomaya A, Jie W (2015) Remote sensing big data computing: challenges and opportunities. Futur Gener Comput Syst 51:47–60. https://doi.org/10.1016/j.future.2014.10.029
26. Zikopoulos P, Eaton C (2011) Understanding big data: analytics for enterprise class Hadoop and streaming data, 1st edn. McGraw-Hill Osborne Media (IBM)
27. NASA (2010) On-orbit satellite servicing study
28. Kambatla K, Kollias G, Kumar V, Grama A (2014) Trends in big data analytics. J Parallel Distrib Comput 74:2561–2573
29. NASA Earth Data (2017) Getting petabytes to people: how the EOSDIS facilitates earth observing data discovery and use. https://earthdata.nasa.gov/getting-petabytes-to-people-how-the-eosdis-facilitates-earth-observing-data-discovery-and-use
30. ITC (2017) ITC-ITC's database of satellites and sensors—all sensors. https://www.itc.nl/Pub/sensordb/AllSensors.aspx

31. Villars RL, Olofson CW, Eastwood M (2011) Big data: what it is and why you should care. White Pap. https://doi.org/10.1080/01616846.2017.1313045
32. Justice CO, Vermote E, Townshend JRG, Defries R, Roy DP, Hall DK, Salomonson VV, Privette JL, Riggs G, Strahler A, Lucht W, Myneni RB, Knyazikhin Y, Running SW, Nemani RR, Zhengming Wan Z, Huete A, van Leeuwen W, Wolfe RE, Giglio L, Muller J, Lewis P, Barnsley MJ (1998) The moderate resolution imaging spectroradiometer (MODIS): land remote sensing for global change research. IEEE Trans Geosci Remote Sens 36:1228–1249. https://doi.org/10.1109/36.701075
33. Datameer (2017) Getting more value from your data lake. https://www.datameer.com/. Accessed 12 Sep 2017
34. Heger D, Ogunleye J (2015) Big data, the cloud and challenges of operationalising big data analytics. Curr Stud Comp Educ Sci Technol 2:427–435
35. Mazhar M, Rathore U, Paul A, Ahmad A, Chen B-W, Huang B, Ji W (2015) Real-time big data analytical architecture for remote sensing application. IEEE J Sel Top Appl Earth Obs, Remote Sens, p 8
36. Datameer (2017) Best practice for a successful Big Data jouney
37. Freitas RM (2011) Virtual laboratory of remote sensing time series: visualization of MODIS EVI2 data set over South America. J Comput Interdiscip Sci 2:57–68. https://doi.org/10.6062/jcis.2011.02.01.0032
38. Vatsavay R, Chandola V (2016) Guest editorial: big spatial data. Geoinformatica. https://doi.org/10.1007/s10707-016-0269-7
39. Zicari RV, Rosselli M, Ivanov T, Korfiatis N, Tolle K, Niemann R, Reichenbach C (2016) Setting up a big data project: challenges, opportunities, technologies and optimization. In: Big data optimization: recent developments and challenges. Studies in big data. https://doi.org/10.1007/978-3-319-30265-2_2
40. González SM, Berbel T dos RL (2014) Considering unstructure data for OLAP: a feasability study using a systematic review. Rev Sist Informação da FSMA 14:26–35
41. Chen M, Mao S, Liu Y (2014) Big data: a survey. Mob Netw Appl 19:171–209. https://doi.org/10.1007/s11036-013-0489-0
42. Khan N, Yaqoob I, Abaker I, Hashem T (2014) Big data: survey, technologies, opportunities, and challenges. Sci World J 18
43. Lang S (2008) Object-based image analysis for remote sensing applications: modeling reality—dealing with complexity. In: Blaschke T, Lang S, Hay GJ (eds) Object based image anal. Springer, pp 3–27
44. Hay GJ, Castilla G (2006) Object-based image analysis: strengths, weaknesses, opportunities and threats (SWOT). OBIA, Int Arch Photogramm Remote Sens Spat Inf Sci 3
45. Blaschke T (2010) Object based image analysis for remote sensing. ISPRS J Photogramm Remote Sens 65:2–16
46. Audubon, Cornell Lab of Orithnology (2017) About eBird|eBird. http://ebird.org/content/ebird/about/
47. Wood C, Sullivan B, Iliff M, Fink D, Kelling S (2011) eBird: engaging birders in science and conservation. PLoS Biol 9
48. Fink D, Hochachka WM, Zuckerberg B, Winkler DW, Shaby B, Munson MA, Hooker G, Riedewald M, Sheldon D, Kelling S (2010) Spatiotemporal exploratory models for broad-scale survey data. Ecol Appl 20:2131–2147. https://doi.org/10.1890/09-1340.1
49. Beddington JR, Agnew DJ, Clark CW (2007) Current problems in the management of marine fisheries. Science 80(316):1713–1716
50. Gorospe KD, Michaels W, Pomeroy R, Elvidge C, Lynch P, Wongbusarakum S, Brainard RE (2016) The mobilization of science and technology fisheries innovations towards an ecosystem approach to fisheries management in the Coral Triangle and Southeast Asia. Mar Policy 74:143–152. https://doi.org/10.1016/j.marpol.2016.09.014
51. Yamaguchi T, Asanuma I, Park JG, Mackin KJ, Mittleman J (2016) Estimation of vessel traffic density from Suomi NPP VIIRS day/night band. Ocean 2016 MTS/IEEE Monterey. OCE 2016:5–9. https://doi.org/10.1109/OCEANS.2016.7761309

52. Straka WC, Seaman CJ, Baugh K, Cole K, Stevens E, Miller SD (2015) Utilization of the suomi national polar-orbiting partnership (npp) visible infrared imaging radiometer suite (viirs) day/night band for arctic ship tracking and fisheries management. Remote Sens 7:971–989. https://doi.org/10.3390/rs70100971
53. Addo KA (2010) Urban and peri-urban agriculture in developing countries studied using remote sensing and in situ methods. Remote Sens 2:497–513. https://doi.org/10.3390/rs2020497
54. Stefanov WL (2001) Monitoring urban land cover change: An expert system approach to land cover classification of semiarid to urban centers. Remote Sens Environ 77:173–185. https://doi.org/10.1016/S0034-4257(01)00204-8
55. Yuliang Q, Buzhou M, Jiuliang F (2000) Study on monitoring farmland by using remote sensing and GIS in Shanxi China. Adv Space Res 26:1059–1064. https://doi.org/10.1016/S0273-1177(99)01118-7

Chapter 3
Structure and Dynamics of Many-Particle Systems: Big Data Sets and Data Analysis

Wolfram Schommers

Abstract In this chapter it is underlined that the structure and dynamics of many-particle systems determine essentially the properties of the systems in solid state physics, materials science and nano-technology. The fundamental description of such systems is based on atoms or molecules that interact with each other. The physical background is discussed in detail. In particular, it is argued that the characteristic features and properties of such systems are already reflected by relatively small parts consisting of 10^2–10^7 particles. It is outlined that for all the systems used in solid state physics etc. no reliable analytical models exist and we have to recourse to numerical methods. The procedure is to solve Newton's equations of motion numerically using the interaction potential as input, and these fundamental equations of motion are expressed by coupled differential equations (molecular dynamics). The essential features of the molecular dynamics method have been discussed. The basic information is very large and we get in such investigations "big data sets", and this information is normally used for data analysis. For the production of the big data sets and their analysis sufficiently large and fast computers are necessary. Does nature also produce big data sets for its operations? This question is discussed critically.

Keywords Many-particle systems · Molecular dynamics · Predictor-corrector algorithm · Data reduction · Data classification · Nano-systems
Basic reality

3.1 Introduction

The fundamental description of systems in solid state physics, materials science and nano-technology is based on many-particle systems where the particles are made of atoms or molecules that interact with each other. The structure and dynamics of the

W. Schommers (✉)
University of Texas at Arlington, Arlington, TX 76019, USA
e-mail: gsw.schommers@t-online.de

N. Dey et al. (eds.), *Big Data for Remote Sensing: Visualization, Analysis and Interpretation*, https://doi.org/10.1007/978-3-319-89923-7_3

particles determine essentially the properties of such systems. Due to the interaction between the atoms or molecules the many-particle systems form a unified whole, and the properties of such systems cannot be considered as a superposition of atom/molecule properties, but there are strong interrelations between them. The number of particles that reflect the characteristics of such man-particle systems is essential to know. How many particles have to be used for the description?

Macroscopic systems normally consist of a large number of particles, and we know that approximately 10^{23} atoms or molecules form the systems in solid state physics and materials science. However, the characteristic features and properties are already reflected through a number of particles that is distinctly smaller than 10^{23}, and the essential characteristics are already reflected within relatively small parts with only 10^2–10^7 particles. It must be investigated carefully what particle number is adequate for the description of a certain material. This is however not necessary when we work within the realm of nano-science and nano-technology, and this is because in such cases the real systems normally consist of a few hundred (thousand) atoms or molecules, depending on the complexity of the system under investigation.

Such kind of many-particle systems describe matter at the smallest level at which functional matter can exist: It is the ultimate level where the properties of materials emerge. But most basic biological structures such as DNA, enzymes and proteins also work at this scale, in particular at the nano-scale, building up, molecule by molecule, macroscopic biological systems we call trees, humans, etc., with their typical intimate features.

For all the systems used in solid state physics etc. no reliable analytical models, expressed through formulas using simplifying models, exist at the ultimate level, and we have to recourse to numerical methods. That is, we have to formulate the basic formulas in the most general form and must solve the problem numerically. The structure and dynamics of the atoms/molecules determine the properties essentially.

The most relevant material properties behave in accordance with classical mechanics, i.e., with Newton's formulas. However, in the formulation of the interaction the quantum-mechanical features of electrons have to be considered carefully. The procedure is to solve Newton's equations numerically using the interaction law as input, and these fundamental equations of motion are expressed by coupled differential equations. This procedure is known in literature as molecular dynamics. In almost all cases the predictor-corrector technique is used as algorithm for the classical equations. Specific predictor-corrector algorithms are quoted below.

The basic information of 10^2–10^7 particles is very large and we get in such investigations "big data sets". If the many-particle system is treated as classical unit, we obtain in realistic simulations with 10^2–10^7 particles for each particle at time τ the position and the momentum, that is, we have

$$\begin{aligned} \tau\colon \mathbf{r}_i &= (x_i, y_1, z_i) \\ \mathbf{p}_i &= (p_{xi}, p_{yi}, p_{zi}); \quad i = 1, 2, \ldots, N \end{aligned} \tag{3.1}$$

where N is the particle number. This leads to big data sets in the case of 10^2–10^7 for N.

This big data set can only be used in specific cases but not for the general analysis of the many-particle system [1]. We have to produce measurable quantities, which consist of reduced data sets. However, for the calculation of the measurable quantities (reduced data sets) the big data set is definitely needed.

For the production of the big data sets and their analysis sufficiently large and fast computers are necessary. Excellently working supercomputers (with GigaMove facilities) are installed for example in Germany in Jülich, Karlsruhe and Munich.

In this chapter we will discuss the basic tools for the treatment of many-particle systems, which are given by the physical laws and algorithms for the production of the big data sets. After that we will talk about analyzing methods that are based on statistical physics, and we will bring examples. In the last section we discuss the question whether nature (reality outside) also uses big data sets for the production of properties that we measure and observe directly in everyday life.

3.2 The Physical Formulation

3.2.1 The Hamilton Function

For the classical description of the many-particle systems in basic form the Hamilton function has to be formulated that is given in the case of monatomic systems by [1]

$$H = \sum_{i=1}^{N} \frac{p_i^2}{2m} + \frac{1}{2} \sum_{\substack{i,j=1 \\ i \neq j}}^{N} \upsilon_{ij} + \frac{1}{6} \sum_{\substack{i,j,k=1 \\ i \neq j \neq k}}^{N} \upsilon_{ijk} + \cdots, \tag{3.2}$$

m is the atomic mass and p_i the momentum of the ith atom. If the electronic arrangement (core electrons) around each ion or atom is sufficiently small the potential energy V may be expanded in terms of many-body interaction potentials and this is reflected through the two interaction terms in Eq. (3.2). The term with υ_{ij} represents the pair potential contribution, and the term with υ_{ijk} reflects the triplet potential term. Higher order contributions are not explicitly quoted in (3.2).

If the polarization of the core electrons is negligible small, then—compared with the pair terms—the higher terms can be neglected and we obtain the pair potential formulation. we normally have:

$$\sum_{\substack{i,j=1\\ i\neq j}}^{N} \upsilon_{ij} = \sum_{\substack{i,j=1\\ i\neq j}}^{N} \upsilon(|\mathbf{r}_i - \mathbf{r}_j|) \tag{3.3}$$

$\mathbf{r}_1, \ldots, \mathbf{r}_N$ are the positions of the N atoms with $|\mathbf{r}_i - \mathbf{r}_j| = r_{ij}$.

Referring to 100 atoms, with (3.3) the problem is significantly reduced (we obtain a 9900-fold sum of values): We have now one pair potential function υ_{ij} with only one dimension (the distance of two particles). This simplification makes it possible to treat many-particle problems up to millions of particles. However, such a far-reaching simplification will influence the applicability [2].

Since we assumed that the atoms behave classically, which is fulfilled in almost all cases and, furthermore, if we assume that only the pair interaction is effective, we obtain the classical Hamilton function in the pair potential approximation, which is given by for monatomic systems by

$$H = \sum_{i=1}^{N} \frac{p_i^2}{2m} + \frac{1}{2} \sum_{\substack{i,j=1\\ i\neq j}}^{N} \upsilon(r_{ij}) \tag{3.4}$$

If three or higher body forces are effective, instead of the pair potential approximation for the potential energy the expansion (3.2) has to be used. Pair potentials and many-body forces are discussed in detail in literature (see, for example [1]).

3.2.2 Simple Models

The properties of a many-particle system is generally formulated on the basis of a given Hamiltonian? If we want to treat the problem conventionally we need simple models. Simple models can be used if controlled approximations are possible, starting from the general case (3.4). In most cases the system are so complex that controlled approximations are not possible, and simple models can only be used for an easy treatment of the many-particle system. However, one should avoid such a way; a realistic description needs more than that.

In solid state physics a simple model can be defined if the so-called harmonic approximation works and if the structure of the solid is crystalline. On this basis, the properties of a lot of materials can be determined successfully. However, this simple model is not always applicable. For example, in the case of the solid electrolyte α-*AgI* the silver subsystem is disordered and behaves anharmonically [3–5]; for this and similar materials simple models are not available.

Also in the case of liquids and gases "simple models" could not be formulated. Even in the case of gases with low densities serious problems are unavoidable.

The reason is simple: In the calculation of the pressure the first terms in the virial expansion are not sufficient, and this is because the expansion converges slowly and, therefore, one has to consider more than the first two terms [6]. The virial coefficients can be expressed in terms of the pair potential $\upsilon(r)$, and the expressions for the higher order virial coefficients become complicated and are not accessible in practical calculations In other words, also in the case of gases "simple model" cannot be formulated.

In summary, the treatment of many-particle systems by "simple models" is hardly possible. For solids in the harmonic approximation and with crystalline structure simple models are available, but in most cases the anharnonicities are relatively strong, even at low temperatures, and anharmonic effects cannot be considered as small perturbations to the harmonic formulation. This is in particular the case for nano-systems because of surface effects. The particles near the surface are less bonded than the particles in the bulk state. Surface particles behave strongly anharmonically The effects can be very large. For example, the melting temperature of typical nano-clusters is distinctly smaller than in the bulk-state of the same material.

For the precise theoretical description of many-particle systems (nano-systems, gases and other specific solids) simple models in the above sense are not relevant.

3.3 Molecular Dynamics

The relatively strong anharmonicities in nano-systems and a lot of other systems are not negligible even at low temperatures. The following question arise: How can the structure and dynamics of nano-systems or similar systems be described? What method or basic model is adequate? Since only for the ideal solid such a basic model exists [other simplifying principles at the microscopic level could not be found for many-particle systems (see Sect. 3.2.2)]; such systems have to be treated with the most general tool, and this is given by the molecular dynamics method. We will give here relevant details about the molecular dynamics method.

But there are other possibilities as, for example, the Monte Carlo approach: In contrast to molecular dynamics with its completely deterministic equations, the Monte Carlo method uses stochastic elements and is a numerical approach as well. The molecular dynamics method enables to study static and dynamic properties, the Monte Carlo method only allows to calculate static properties; within the Monte Carlo method configurational averages and thermodynamic quantities can be determined. Also, quantum molecular dynamics is a suitable method.

In conclusion, the method of molecular dynamics allows the most general description for the determination of the properties. Such calculations have to be performed for the characterization of many-particle systems. If a system behaves classically, molecular dynamics are done in most cases on the basis of (3.4), that is, if the pair potential approximations can be used and higher-order many-body interactions can be neglected. In the following we would like to give more details concerning the molecular dynamics method in connection with "big data sets".

3.3.1 Basic Information

For the general determination of the structure and dynamics of systems with strong anharmonicities the molecular dynamics method is important, because the anharmonicities can be treated without approximation. Here simple models, mostly used in connection with uncontrolled simplifications, are not needed and not known so far; phenomenological or empirical models for the treatment of many-particle systems have to be avoided. As we already mentioned several times, the main reason is that the typical many-particle systems behave strongly anharmonically and the phonon picture is not applicable. The molecular dynamics method allows to consider all kind of anharmonicities without approximation. What is here the procedure for the solution of the many-particle system? We start with Hamilton's equations

$$
\begin{aligned}
\dot{p}_{x_i} &= -\frac{dH}{dx_1}, \ldots, \dot{p}_{z_N} = -\frac{dH}{dz_N} \\
\dot{x}_1 &= \frac{dH}{dp_{x_i}}, \ldots, \dot{z}_N = \frac{dH}{dp_{z_{Ni}}}
\end{aligned}
\tag{3.5}
$$

which are solved numerically in an iterative way. This is done with the help of a high-speed computer, and we obtain the following set of information for N atoms:

$$
\begin{aligned}
&q(\tau_1), p(\tau_1) \\
&q(\tau_2), p(\tau_2) \\
&\vdots \\
&q(\tau_i), p(\tau_i) \\
&\vdots
\end{aligned}
\tag{3.6}
$$

where

$$
\begin{aligned}
q(\tau_i) &= (x_1, y_1, z_1, \ldots, x_N, y_N, z_N), \\
p(\tau_i) &= (p_{x_1}, p_{y_1}, p_{z_1}, \ldots, p_{x_N}, p_{y_N}, p_{z_N}).
\end{aligned}
\tag{3.7}
$$

In other words, as solution of the classical laws we obtain the coordinates and momenta for all particles of the system as a function of time τ. The time step in the iteration process has to be estimated and is dependent on the system under investigation. We do not give the details in this chapter. The solution (3.6) represents the

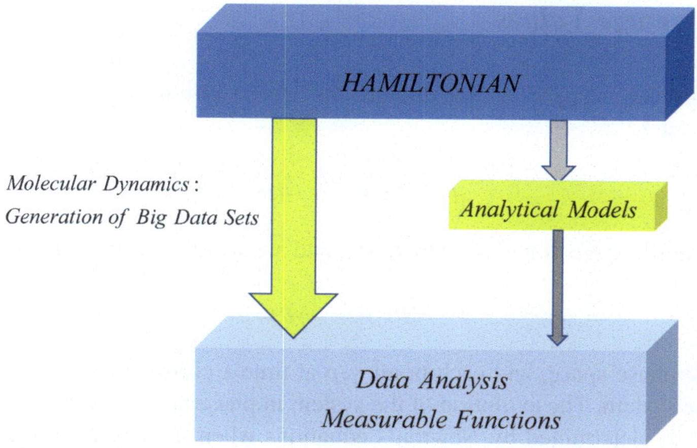

Fig. 3.1 The properties of many-particle systems (solids, liquids, nano-systems, etc.) can be determined on the basis of simple models (analytical models); in most cases additional conditions have to be applied. In contrast to this way, the molecular dynamics method allows the description without simple models and additional assumptions. This is an advantage and important since in most cases such systems are so complex that simple models with analytical solutions cannot be formulated without the use of uncontrolled simplifications starting from the general case (3.4). For nano-systems etc. simple models could not be found up to now. Therefore, the treatment by molecular dynamics is of particular relevance here; the solution (3.6) represents the most general form for the many-particle system and reflects a large amount of data, that is, it is a "big data set"

most general form for the many-particle system and reflects a large amount of data, that is, it is a "big data set".

If the system consists of N particles the information for k iteration steps is given in the time interval

$$\tau_G = k \cdot \Delta\tau \tag{3.8}$$

with

$$\Delta\tau = \tau_{i+1} - \tau_i \tag{3.9}$$

The total information of the many-particle system is expressed by (3.6). This information is the basis for the determination of the properties of the system, also those that are reflected through experimental data. In the following we will give some remarks about the determination of typical functions on the basis of information (3.6) (Fig. 3.1).

3.3.2 Average Values

Let us consider a space of 6N dimensions whose points are expressed by the 3N coordinates

$$q = (q_1, q_2, \ldots, q_{3N}) \tag{3.10}$$

To each coordinate belongs a momentum, and we have the 3N momenta

$$p = (p_1, p_2, \ldots, p_{3N}). \tag{3.11}$$

This is the phase space, and each point q, p at time τ corresponds to a mechanical state of the system. The evolution of the system in phase space as a function of time is completely determined by Newton's equations when we work classically.

The collective motion is described by a trajectory in phase space (see Fig. 3.2a) if the system behaves classically. This trajectory moves through phase space and the points in Fig. 3.2b indicate how often the elements $dqdp$ at point (q, p) have been passed through by the trajectory given in Fig. 3.2a.

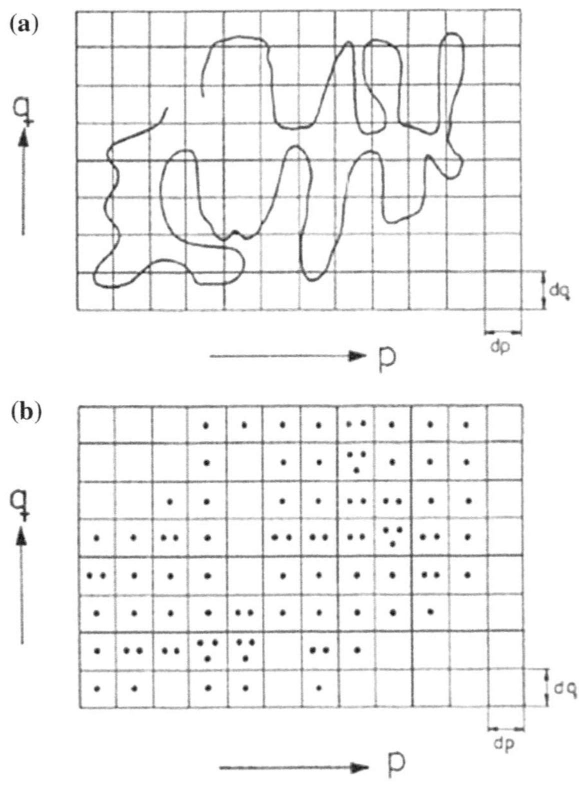

Fig. 3.2 In figure **a**, the trajectory in phase space is shown. Figure **b** reflects a statistical ensemble. Details are given in the text

Thus, we have two representations: The trajectory in Fig. 3.2a and a "cloud" of phase points in Fig. 3.2b. The "cloud" is a big data set and reflects a large number of systems of the same nature; at a given instant they differ with respect to the configurations and momenta. On the one hand, we have a single dynamical system (Fig. 3.2a) and, on the other hand, we have various systems (Fig. 3.2b): All situations follow the classical formulas introduced by Hamilton's equations (3.5). The collection of systems, the so-called statistical ensemble, is the system in one of its possible states. The statistical ensemble is useful when the relationship between dynamics and thermodynamics is considered. In particular, it is a tool for the reduction of the big data set. This is an important point because this way also allows to describe experimental data. In other words, the data reduction performed by this method is reasonable.

3.3.3 Statistical-Mechanical Ensembles

The function

$$\rho(q_1, q_2, \ldots, q_{3N}, p_1, p_2, \ldots, p_{3N}, \tau) \rightarrow \rho(q, p, \tau) \tag{3.12}$$

describes the density of the statistical ensemble in phase space. Since the number of points of the statistical ensemble is arbitrary, the function $\rho(q,p,\tau)$ has to be normalized and we have

$$\int \rho(q,p,\tau)dqdp = 1. \tag{3.13}$$

The quantity

$$\rho(q,p,\tau)dqdp$$

reflects the probability of finding a system of the ensemble at time τ in the element $dqdp$ at the phase space point (q, p).

The density (3.12) has to take an asymptotic value for sufficiently large times τ:

$$\lim_{\tau\to\infty} \rho(q,p,\tau) = \rho(q,p) \tag{3.14}$$

Only in this case the density (3.12) is meaningful. The following condition is relevant: When we observe the trajectory $q(\tau), p(\tau)$ over a sufficiently long period of time, the function $\rho(q,p)$ should be reproduced (ergodic hypothesis). Then, we have equivalence between the average over the trajectory and the average over the statistical ensemble. In the case of the function $f[q(\tau), p(\tau)]$ we get

$$\langle f \rangle = \frac{\int f(q,p)\rho(q,p)dqdp}{\int \rho(q,P)dqdp} \tag{3.15}$$

$$\langle f \rangle = \lim_{\tau \to \infty} \frac{1}{\tau} \int_0^{\tau} f(\tau)d\tau \tag{3.16}$$

For the density function $\rho(q,p)$ different expressions exist that are extensively discussed in literature. These expressions are dependent on the thermodynamics of the outside world in which the system under investigation is embedded. Let us list here the results for the function $\rho(q,p)$ for the three most relevant situations; the details are not discussed.

Microcanonical ensemble:
In this case we have a separated system (the parameters N, V and E are given, where E is the energy), and we have

$$\rho(q \cdot p) = \begin{cases} \rho_0 = const \\ 0 \quad otherwise \end{cases} \tag{3.17}$$

for

$$E < H(q,p) < E + \Delta \tag{3.18}$$

where $E < \Delta$. With (3.15) we obtain

$$f = \frac{\int_{E<H(q,P)<e+\Delta} f(q,p)dqdp}{\int_{E<H(q,P)<E+\Delta} dqdp} \tag{3.19}$$

Canonical ensemble:
The system under investigation is treated as a closed, isothermal system (the parameters N, V and the temperature T are given), and the density function $\rho(q,p)$ is given in this case through

$$\rho(q.p) = \frac{\exp\left\{-\frac{H(q,p)}{k_B T}\right\}}{\int \exp\left\{-\frac{H(q,p)}{k_B T}\right\} dqdp} \tag{3.20}$$

Grand canonical ensemble:
In this case the parameters V, T and μ are given, where μ is the chemical potential, and the density function $\rho(q,p)$ is expressed by the following relation:

$$\rho(q.p) = \frac{\exp\left\{-\frac{H(q,p)-\mu N}{k_B T}\right\}}{\int \exp\left\{-\frac{H(q,p)-\mu N}{k_B T}\right\} dqdp} \tag{3.21}$$

Thermodynamic Limit:
The thermodynamic limit is defined through

$$\begin{aligned} &N \to \infty \\ &V \to \infty \\ &\tfrac{N}{V} = const \end{aligned} \tag{3.22}$$

and the three ensemble types, i.e., the microcanonical ensemble, the canonical ensemble and the grand canonical ensemble, are equivalent, i.e., within the thermodynamic limit we obtain in all three cases the same value for the statistical average $\langle f \rangle$ of $f(q,p)$.

3.4 Individual and Mean Velocities

The statistical ensemble (Fig. 3.2b) allows to define averages but also in connection with the trajectory (Fig. 3.2a). Again, the system under investigation can take all states that are given by the statistical ensemble. We may also consider the trajectory given in Fig. 3.2a; in this case the system moves through phase space in the course of time τ.

Let us consider a single particle of a monatomic system consisting of N particles, and let $v(\tau)$ be its velocity. Then, we can express the probability dW of finding the velocity of the single particle in the interval $v, v+dv$, and this probability is expressed by Maxwell's distribution as follows:

$$g(v) = \frac{4}{\sqrt{\pi}}\left(\frac{m}{2k_B T}\right)^{3/2} v^2 \exp\left\{-\frac{mv^2}{2k_B T}\right\}, \tag{3.23}$$

Using this $g(v)$ the probability is given by $dW = g(v)dv$. In Eq. (3.23) m is the mass of the particle, T is the temperature of the system under investigation and k_B expresses Boltzmann's constant.

Then, the mean-square velocity $\langle v^2 \rangle$ is given by the following procedure: We have to average v^2 over all states (members) of the ensemble, and these states for a single particle are given by the cloud (Fig. 3.2b is an example). Using Maxwell's distribution (3.23) we immediately obtain for the mean-square velocity

$$\langle v^2 \rangle = \frac{4}{\sqrt{\pi}} \left(\frac{m}{2k_B T} \right)^{3/2} \int_0^{\infty} v^4 \exp\left\{ -\frac{mv^2}{2k_B T} \right\} dv = 3\frac{k_B T}{m} \tag{3.24}$$

Exactly the same value for $\langle v^2 \rangle$ will be obtained if we average v^2 over the trajectory (Fig. 3.2a), that is, we average $v^2 = v^2(\tau)$ over the states v^2 which the single particle takes in the course of time τ. Let us define a function $v^2(\tau_0)$ by

$$v^2(\tau_0) = \frac{1}{\tau_0} \int_0^{\tau_0} v^2(\tau)\, d\tau \tag{3.25}$$

This function must have the following property:

$$\lim_{\tau_0 \to \infty} v^2(\tau_0) = \langle v^2 \rangle \tag{3.26}$$

This exactly means that for a sufficiently large time τ_0 the velocity $v(\tau)$ of a particle of the ensemble (N-particle system) has occupied all the states expressed through Maxwell's distribution (3.23). It turned out within realistic molecular dynamics calculations the condition $\tau_0 \to \infty$ in (3.26) is already fulfilled after a relatively short time; a typical example is $\tau_0 \approx 10^{-11}$ s. This a relatively large time at the atomic scale.

3.4.1 Initial Values

Again, from the solution of the classical equations we obtain as a result all coordinates and all momenta (velocities) of the N particles as a function of time τ. This is the complete atomic information about the system consisting of N particles. However, for the solution of the differential equations initial values for all the coordinates and all the velocities of the N particles are needed. Let us give a few remarks on this point.

Initial State for the Coordinates:
The particles can be distributed randomly if we work with liquids or gases. However, we have to choose an appropriate density. In the case of crystals, the particles are normally positioned within an array so that the perfect lattice structure, appropriate to the system under investigation, is generated; this is valid for crystals with and without surface and of course also for systems at the nano-scale. The system under investigation is not fixed to this initial structure, but the structural configuration develops in the course of time until a stationary state with a realistic structure is reached.

Initial States for the Velocities (Momenta):
The directions of the velocities

$$\mathbf{v}_i/|\mathbf{v}_i|, \quad i = 1, 2, \ldots, N, \tag{3.27}$$

should be distributed randomly at the beginning of the calculation. so that the sum over

$$\sum_{i=1}^{N} \mathbf{v}_i/|\mathbf{v}_i| = 0 \tag{3.28}$$

must be zero (or constant). This is the condition when there are no external forces acting on the system. Equation (3.28) must be valid for all times τ; the condition (3.28) is necessary because the conservation of momentum must be fulfilled at each time τ, that is, during the molecular dynamics calculation.

If the system is in thermal equilibrium, the magnitudes of the particles follow Maxwell's distribution. We can however choose for all particles the same magnitude for the velocities as initial values. This is a more convenient way instead to use Maxwell's distribution as initial input. However, the distribution for the velocities develops in the course of time towards Maxwell's distribution (3.23), and this is due to the mutual interaction between the N particles.

In conclusion, when we choose the same magnitude for the velocities as initial values, the initial probability $dW = g(v)dv$ of finding the velocities between $v, v + dv$ is exactly given by a delta function. In other words. the system, which we investigate in this way, is initially not in thermal equilibrium. But we can study the behavior of the velocity distribution by means of the function

$$\alpha(\tau) = \frac{\frac{1}{N}\sum_{i=1}^{N}\left[v(\tau)_i^2\right]^2}{\left[\frac{1}{N}\sum_{i=1}^{N} v(\tau)_i^2\right]^2} \tag{3.29}$$

where the velocities $\mathbf{v}_i(\tau)$, $i = 1, 2, \ldots, N$, which appear in (3.29), are again the calculated velocities by the molecular dynamics calculation. With (3.29) we can easily study when Maxwell's distribution is reached. The function $\alpha(\tau)$ takes exactly the value 5/3 for all times τ when Maxwell's distribution is fulfilled. When we choose the same magnitude for the velocities as initial values for the initial velocity distribution, we get $\alpha(\tau) = 1$.

Normally, that thermal equilibrium $(\alpha(\tau) = 5/3)$ is reached after a short time period. It is a matter of experience that thermal equilibrium is reached after a few hundred time steps when we start from $\alpha(\tau) = 1$; the time step is small and is of the order of 10^{-14} s. Due to the finite number of particles, the system fluctuates around 5/3 (see also Fig. 3.3); these fluctuations decrease with increasing particle number N and are physically realistic, i.e., also in nature we would have such fluctuations. Clearly, in the case of sufficiently large, macroscopic systems $(N \rightarrow \infty)$ such

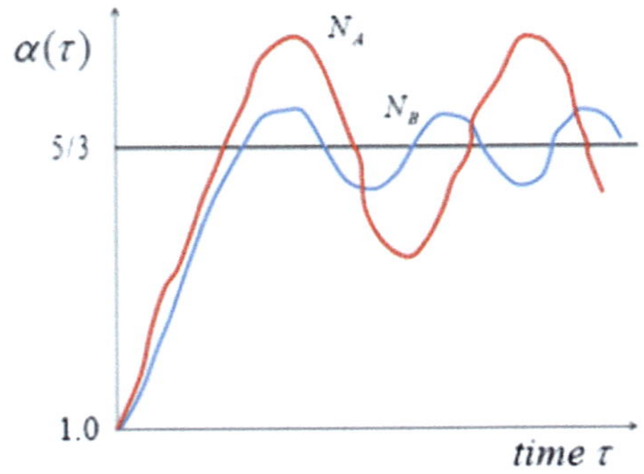

Fig. 3.3 The behaviour of the function $\alpha(\tau)$ in the course of time τ (schematic representation). Initially, the function $\alpha(\tau)$ takes the value 1: $\alpha(\tau) = 1$, i.e., the system is initially not in thermal equilibrium. However, Maxwell's distribution is reached after a relatively short time period; normally only a few hundred molecular dynamical time steps are needed in order to get thermal equilibrium. Due to the finite number of particles, the system fluctuates around 5/3. These fluctuations increase with decreasing particle number, that is, we have $N_A < N_B$. The function $\alpha(\tau)$ is independent of the interaction potential used in the calculation and is also independent of the mass. So, $\alpha(\tau)$ fluctuates around 5/3 for every system

fluctuations disappear and $\alpha(\tau)$ is a constant and independent of time τ. The function $\alpha(\tau)$ is independent of the interaction potential used in the molecular dynamics calculation.

Temperature Behaviour of Molecular Dynamics Systems:

The temperature of systems with a finite particle number, like molecular dynamics systems, also fluctuates and is dependent on time τ. The temperature is defined by the well-known relation

$$\frac{1}{2}m\left\langle \mathbf{v}(\tau)^2 \right\rangle = \frac{3}{2}k_B T(\tau) \tag{3.30}$$

with

$$\left\langle \mathbf{v}(\tau)^2 \right\rangle = \frac{1}{N}\sum_{i=1}^{N} \mathbf{v}_i^2(\tau) \tag{3.31}$$

we obtain for the temperature

$$T(\tau) = \frac{m}{3Nk_B}\sum_{i=1}^{N} \mathbf{v}_i^2(\tau) \tag{3.32}$$

These temperature fluctuations (Fig. 3.4) do not reflect an artificial effect but actually appear in nature when we work with small units, such as nano-systems.

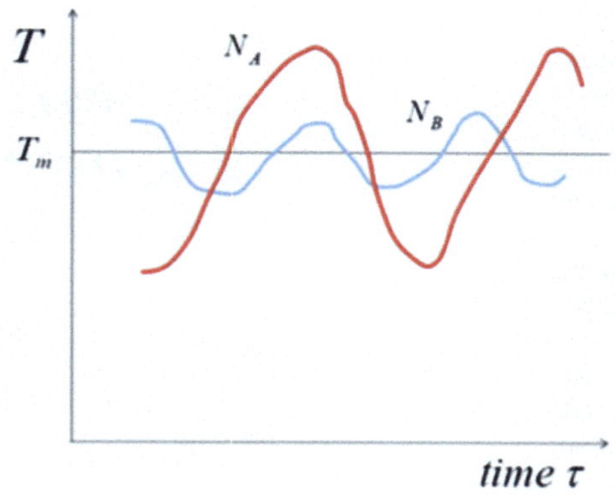

Fig. 3.4 The temperature for molecular dynamics systems vary with time τ; here schematically represented. The temperature fluctuates around the mean temperature T_m, and this is again due to the finite number of particles. The temperature fluctuations increase with decreasing particle number, that is, we have $N_A < N_B$

In other words, temperature fluctuations are quite natural. Such fluctuations are reflected in the material properties; the specific heat at constant volume is an example. The details are not of interest in connection with this chapter.

3.4.2 Measureable Quantities

Again, the coordinates and momenta for all particles as a function of time is the complete information about a microscopic classical many-particle system, and this is just the information we obtain from molecular dynamics.

However, only certain averaged quantities can be measured, i.e., this total information is not directly accessible to measurements. Just those quantities that are characteristic for the structure and dynamics of many-particle systems (nano-systems, systems with and without surface), In [1] we have discussed the properties of clusters made of aluminium and other essential structures.

For the comparison with measurements, averaged quantities are relevant, and the theoretical expressions have to reflect such kind of averages, which can be calculated on the basis of the big data set, that is, by means of the fundamental molecular dynamics information (the coordinates and velocities of all N particles as a function of time). The pair correlation function, the triplet correlation function and the (generalized) phonon density of states are typical examples. The situation is schematically shown in Fig. 3.5: From molecular dynamics and big data sets to data analysis.

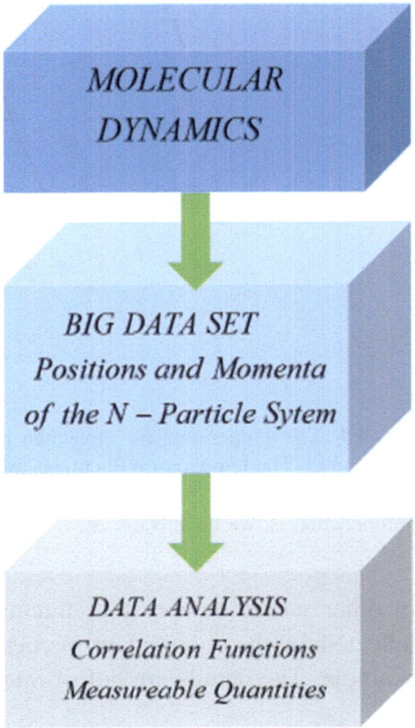

Fig. 3.5 The complete information is given by the coordinates and the momenta (velocities) of all N particles as a function of time. This is the complete information and represents a big data set expressed by (3.6). From this big data set we can directly calculate measurable quantities (data analysis). as, for example, the generalized phonon density of states and structural data (pair correlation function, which is accessible to measurements)

3.5 Correlation Functions

An adequate analysis of many-particle systems, just in connection with measurements, correlation functions are of particular interest. Here correlation functions of the type

$$\langle a(\tau')b(\tau'')\rangle \tag{3.33}$$

are in the focus, where the elements $a(\tau')$ and $b(\tau'')$ are formulated through

$$\begin{aligned} a(\tau') &= a(q(\tau'), p(\tau')), \\ b(\tau'') &= b(q(\tau''), p(\tau'')). \end{aligned} \tag{3.34}$$

The brackets $\langle \cdots \rangle$ in (3.33) denote thermodynamic averaging, which has been discussed in connection with molecular dynamics calculations and the various ensembles. In statistical mechanics the time evolution of systems is expressed by the operator $\hat{S}(\tau)$

$$a(q(\tau'), p(\tau')) = \hat{S}(\tau' - \tau'')a(q(\tau''), p(\tau'')) \tag{3.35}$$

where $\hat{S}(t)$ can be formulated by means of the well known Liouville operator $\hat{L}$ as follows

$$\hat{S}(\tau) = \exp(i\hat{L}\tau) \tag{3.36}$$

with

$$\hat{L} = i\sum_{i=1}^{3N}\left[\frac{\partial H}{\partial q_i}\frac{\partial}{\partial p_i} - \frac{\partial H}{\partial p_i}\frac{\partial}{\partial q_i}\right], \tag{3.37}$$

where q_i and p_i denote again all the 3*N* coordinates and momenta of the system.

In general, it is difficult to determine the time evolution of many-particle systems on the basis of (3.36). However, on the basis of the big data set, i.e., the molecular dynamics information (3.6), it is straightforward to determine correlation functions of type (3.33). However, we will not deepen this point in this chapter, but we must refer to the literature.

In connection with time correlation functions of type (3.33) it is important to mention that the time origin is not relevant, i.e., for the correlations only time differences have to be considered. In other words, in (3.33) not τ' and τ'' are relevant but τ''–τ':

$$\langle a(\tau')b(\tau'')\rangle = \langle a(0)b(\tau'' - \tau')\rangle. \tag{3.38}$$

Let us briefly discuss an example: With $a(0) = \mathbf{v}(0)$ and $b(\tau) = \mathbf{v}(\tau)$ we get the velocity autocorrelation function

$$\phi(t) = \langle \mathbf{v}(0) \cdot \mathbf{v}(\tau)\rangle, \tag{3.39}$$

where $\mathbf{v}(\tau)$ is the velocity for one atom of the statistical ensemble at time τ. With $\phi(\tau) = \phi(-\tau)$ the Fourier transform of $\phi(\tau)$ is given by

$$f(\omega) = \frac{2}{\langle \mathbf{v}(0)^2\rangle\pi}\int_0^\infty \phi(\tau)\cos\omega\tau\, d\tau, \tag{3.40}$$

where $f(\omega)$ fulfills the norm:

$$\int_0^\infty f(\omega)d\omega = 1. \tag{3.41}$$

In the case of the harmonic crystal the spectrum $f(\omega)$ just describes the frequency spectrum of the normal modes that are known as phonons. The frequency spectrum $f(\omega)$ defined by (3.40) is quite general and is applicable to systems with strong anharmonicities like nano-systems. $f(\omega)$ describes the complete dynamics of many-particle systems, that is, all kinds of excitations are included—even diffusion processes. The diffusion constant is directly expressed by $f(\omega)$ [7]:

$$D = k_B T\pi/(2m)f(\omega = 0). \tag{3.42}$$

In particular, using $f(\omega)$ in the description of the dynamics no problems appear in connection with systems with surfaces where the dynamical matrix solutions for the calculation of the modes are obviously not complete as we have demonstrated in [1].

How can we calculate the velocity autocorrelation function $\phi(\tau)$) and other properties from the basic molecular dynamics information (3.6)? In order to answer this question, we have to outline how thermodynamic averages $\langle\cdots\rangle$ can be calculated when we work within molecular dynamics, which we have introduced in the context of (3.33). Within molecular dynamics calculations averages will not be calculated on the basis of a statistical ensemble. The calculated information expressed by (3.6) already represent the data in thermal equilibrium. Therefore, it is straightforward to determine averages on the basis of (3.6). For the mean-square velocity

$$\phi(0) = \langle \mathbf{v}(0)^2\rangle = \langle \mathbf{v}^2\rangle \tag{3.43}$$

[see (3.39)] the molecular dynamical average is simply given by

$$\langle \mathbf{v}^2\rangle = \frac{1}{N_\Theta}\frac{1}{N}\sum_{j=1}^{N_\theta}\sum_{i=i}^{N}\mathbf{v}_i^2(\theta_j). \tag{3.44}$$

The average has to be formed over all N particles and various times θ_j, where N_Θ is the number of initial times.

3.6 Potential Functions

In this section, we would like to give some general remarks on potential functions. The interaction is an essential input in the performance of molecular dynamics calculations and cannot be done without it. The interaction potential must be reliable when we try to describe the systems realistically. Here we want to quote the types of interaction, which are relevant in condensed matter physics. How sensitive are the properties of many-particle systems against potential-variations? We will answer this question, but only very shortly. Metallic nano-systems are of particular interest, and we will discuss aluminium as a typical example.

3.6.1 Types of Interactions in Condensed Matter Physics

Basically, in condensed matter physics we have to distinguish between 4 binding types (see, for example, [8]): (1) *Ionic interactions* (4–14 eV), (2) *Metallic bonds* (0.7–6 eV), (3) *Van-der-Waals interactions* (0.02–0.3 eV), (4) *Covalent bonds* (1–10 eV), where the numbers denote the typical binding energies. Also, the *hydrogen bridge bond* (0.1–0.5 eV) is a characteristic binding type but will not be discussed in the following.

Binding between Ions:
Here the many-particle systems consists of different ions (for example, *Na* and *Cl*), i.e., it consists of atoms, which have exchanged electrons and we get positive and negative charged ions of radii r_+ and r_-. The interaction potential is given by Coulomb's law, which is dependent on the distance r between the ions if the distances are larger than $r_+ + r_-$. The interaction is repulsive for distances $r < r_+ + r_-$, and the core electrons overlap each other; and Coulomb's law is no longer applicable in the simple form.

Metallic Systems:
If the electrons of the outer shell are only weakly bonded, the electrons may leave the atom and can move through the whole many-particle system and we obtain a sea of conduction electrons. In this case we have a metal consisting of positively charged ions that are surrounded by the conduction electrons. In this case the ion-ion interaction is screened by the conduction electron leading to a complex situation. Such metallic systems are often treated on the basis of the so-called pseudo-potential theory. Concerning models, in [1] some details are given.

Van-der-Waals Forces:
The electrons move around the atomic nucleus and this means that the atoms have at instant τ an electrical dipol moment. This dipol moment is however in the middle zero when the time average is considered. The atoms of a many-particle system interact with each other via these momentary dipol moments. The calculations show that this interaction is attractive. Typical van-der-Waals systems are systems consisting of noble gas atoms However, also repulsive core effects at small distances are relevant as well.

Covalent Interactions:
Covalent bindings are defined through the fact that a certain part of electrons can belong to several particles at the same instant. The probability of finding the electron, which is responsible for binding, is between the particles, and this effect is responsible for the interaction between them (covalent binding). Here spin saturated electron pairs are relevant, leading to the effect that the electron shells take a noble gas configuration. Carbon, amorphous semiconductors, hydrogen molecules etc. are examples for covalent binding.

Discussion:
The interaction for a specific material has to be constructed carefully since the properties are sensitive to the potential form. Different binding types can be mixed for a certain material. For example, in metals not only the metallic binding is relevant but also van-der-Waals interactions can contribute.

3.6.2 Remarks Concerning Phenomenological Potentials

In connection with realistic calculations, phenomenological potential functions are used very often, which are often not strictly constructed with respect to the basic interaction types discussed above. The potential of Morse [9] is such a phenomenological potential and has often been applied: Within lattice dynamic [10] and also for the study of defect structures in metals [11]. Furthermore, it has been used for the study of inert gases in metals and the equation of state has been investigated as well. Further application in connection with elastic metal properties [12], the interaction between gas atoms and crystal surfaces, and in studies of other specific problems.

For the calculation of the energy levels of diatomic molecules, Morse formulated some conditions for the interatomic potential of the atoms of the molecule [2] in order to describe spectroscopic properties of the molecules. Morse proposed the following model potential:

$$\upsilon(r) = \alpha_0 \exp[-2\alpha(r - r_0)] - 2\alpha_0 \exp[-\alpha(r - r_0)], \tag{3.45}$$

where r_0 is the intermolecular separation. The solution of Schrödinger's wave equation (its radial part) with this potential yielded the correct energy levels.

In conclusion, the potential by Morse has not been constructed on the basic interaction types discussed above, but is based on the spectroscopy of molecules. Nevertheless, the potential has been used very often in the study of various many-particle properties, but there is no deeper physical justification for that.

There are many other phenomenological approaches for the interaction (for example, the Buckingham potential), but we do not want to discuss these functions here because in most cases there is no or almost no physical background for their introduction; these potentials have often been chosen for purely pragmatic reasons and, therefore, they have to be considered only as fitting functions. The Lennard-jones potential is often used in the analysis of many-particle systems, this potential is given by [1]

$$\upsilon(r) = 4\varepsilon\left[\left(\frac{\sigma}{r}\right)^{12} - \left(\frac{\sigma}{r}\right)^{6}\right] \tag{3.46}$$

and is very often used in connection with molecular dynamics, but it turned out that this potential type can only be used for rough estimations, even in the case of noble gas systems.

Since there is a strong potential-sensitivity in connection with the properties of systems with surface (for example, nano-systems), we have to construct the potentials very carefully. In [1] we discussed some relevant theoretical methods for the construction of model potentials, in particular also model-independent methods.

As we remarked several times, it turned out that the properties of many-particle systems vary strongly in the case of small potential changes, i.e., the properties are sensitive to variations in the interaction potentials. Therefore, the potentials have to be modeled carefully. What does "sensitive to small variations in the interaction potentials" mean? In [1] a detailed discussion is given.

3.6.3 Metallic Nanosystems

The interaction potentials for nano-systems and other many-particle systems, in particular then when surfaces are involved, have to be constructed realistically, and when we study nano-systems it has to be taken into account that a large number of atoms belongs to the surface region of the system. This effect is particularly significant in the case of metals.

The determination of pair potentials for metals can be done, for example, by means of pseudopotential theory that we have discussed in literature in all details. In this section, we would like to quote some relevant points in connection with a model potential for *Al*, as a typical example for a metal system.

3.6.4 Pair Potential for Al

The temperature-dependence of properties can easily be studied by molecular dynamics. However, in some cases the input, that is, the potential can be temperature-dependent as well. This is not case for noble gases (for example, krypton [13]); here the properties can be studied with a temperature-independent pair interaction. This is however not possible in the case of metals.

We studied *Al* for a realistic set of potentials [14, 15]. Finally, we obtained a pseudopotential with consideration of van-der-Waals-type interactions. This potential has been checked by a lot of experimental data:

(1) Phonon density of states.
(2) The structure and also the dynamics of *Al* in the liquid state.
(3) The melting temperature could be fixed accurately.
(4) We confirmed an interesting experimental finding: For the outermost surface layer an outward relaxation has been observed.

(5) LEED experiments: The calculated mean-square displacements at the *Al*-surface agree well with the experimental data.
(6) Anharmonic effects already at low temperature could be confirmed.
(7) The onset of premelting is described realistic.

In conclusion, the careful determination of the potential functions, which are the input in molecular dynamics calculations, is necessary when we want to describe the properties of many-particle systems realistically.

3.7 Visualization and Algorithms

Molecular dynamics calculations produce big data sets, i.e., a large amount of information is given within the frame of such investigations. Unfortunately, in realistic cases it is not possible to evade that. The reason is simple: Due to the strong anharmonic effects in connection with the interaction potentials, simple models could not be constructed so far in order to make the production of big data sets superfluous.

There is no way out, the big data set must be produced and must be systematically reduced after that, and this is because this large amount of information cannot be ordered in accordance with established principles. However, certain procedures allow the use of big data sets for the calculation of measurable properties, and this is a reduced representation of a big data set. In other words, the physical analysis of the system under investigation leads to a reduced data set in a most natural way. This is an important point and gives big data sets sense.

Nevertheless, visualization methods can be helpful during the design of molecular dynamics models. Huge amounts of data, produced by molecular dynamics given through the positions and velocities in original form, for thousands of particles and thousands of time steps, can be used for the visualization of atomic processes. No doubt, the numerical analysis of the molecular dynamics data on the basis of physical laws is the most relevant task, but it is also instructive to visualize the particles. We don't go in detail here, but a good example in given in the next subsection in connection with wear.

3.7.1 Wear at the Nano-level

In the case of Hamiltonian systems, such as molecular dynamics systems, *friction* as we know it from macroscopic descriptions, is not defined. Friction at the macroscopic level is based on forces that are velocity-dependent. The forces at the microscopic level are dependent on the structure (particle positions) but not on the particle velocities. Therefore, a force, which is proportional to the velocity, cannot

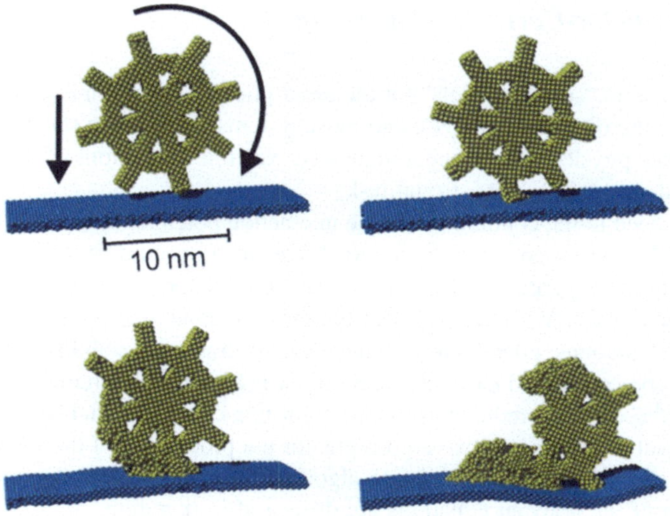

Fig. 3.6 The figure shows a wheel consisting of aluminium atoms. For this configuration molecular dynamics calculations have been performed. The wheel rotates with 10^{12} rev/s and its diameter is approximately 10 nm. The wheel can be destroyed, as can be seen in the figure; this effect is dependent on the magnitude of the force vertically applied. © World Scientific (2004)

be introduced microscopically and this means that a friction constant in the macroscopic sense is not definable within modern materials research in which microscopic molecular dynamics calculations play the basic role.

In the case of non-Hamiltonian systems friction can be studied by non-equilibrium molecular dynamics, but we do not want to deepen this point here.

At the nano-level, wear is important and reflects specific complex processes. The example in Fig. 3.6 illustrates that. In Fig. 3.6 a wheel is shown and it has a diameter of approximately 10 nm. It is a model that is used in a molecular dynamics calculation. This wheel is pressed on a surface. It rotates with 10^{12} revolutions per second, and it has a temperature of 300 K Due to the pressing on the surface, wear effects emerge. The wheel can be destroyed, as can be observed in Fig. 3.6; this effect is dependent on the magnitude of the force vertically applied.

In summary, in the microscopic realm friction (wear) is a complex process and cannot be characterized by only one constant as is usual in macroscopic descriptions. In our example (Fig. 3.6) also the surface structure has a definite influence on wear-effects and the shape of the wheel as well. This example makes clear that during the design and construction of molecular dynamics models the visualization of the particles can be helpful.

3.7.2 *Predictor-Corrector Algorithm*

Since we have no "simple models" for the description of anharmonic many-particle systems (see the discussion above), and most systems in materials science, biology and solid state physics behave anharmonically, Newton's equations of motion have to be solved, and this is not a trivial task.

Here we need initial conditions and the interaction potential as input. What is the solution of Newton's equation of motion? We need specific procedures for the production of the big data sets. Here the so-called predictor-corrector algorithms are of particular interest. We have to select the most suitable algorithm for molecular dynamics calculations; effectiveness, efficiency, accuracy and stability are essential and have to be considered carefully, and only a few algorithms remain. All that is fulfilled in the case of predictor-corrector algorithms. A good analysis is given in [16]. We want to quote only two algorithms for the production of the big data sets: The Verlet algorithm and the Nordsieck algorithm, but a lot of further proposals for the solution of the relevant equations are discussed in literature.

Other numerical integration methods for the solution of differential equations can be found in almost any general textbook on applied mathematics, but are in the most cases not suitable for molecular dynamics. For example, the common Kutta-Runge method and its variants can be ruled out.

3.7.2.1 Verlet Algorithm

The Verlet method [16] is widely used for solving Newton's classical equations of motion. Within the framework of this algorithm the current position $\mathbf{r}_n$ and the current acceleration $\mathbf{a}_n$ are used, but also the previous position $\mathbf{r}_{n-1}$ is needed for the determination of position $\mathbf{r}_{n+1}$ for the next time step. Without going in detail, within Verlet's algorithm the following procedure is used:

$$\mathbf{r}_{n+1} = 2\mathbf{r}_n - \mathbf{r}_{n-1} + \Delta\tau^2\mathbf{a}_n \tag{3.47}$$

with

$$\begin{aligned} \mathbf{r}_n &= \mathbf{r}(\tau_n) \\ \mathbf{a}_n &= \frac{\mathbf{F}(\tau_n)}{m} \\ \tau_n &= n\Delta\tau, \quad n = 0, 1, 2, \ldots \end{aligned} \tag{3.48}$$

F denotes a force; the forces have to be calculated with the help of the interaction potential, which are system-specific. The velocities $\mathbf{v}_n$ are simply given by

$$\mathbf{v}_n = \frac{\mathbf{r}_{n+1} - \mathbf{r}_{n-1}}{2\Delta\tau} \tag{3.49}$$

The basic Verlet algorithm has been successfully improved. The so-called half step leap-frog scheme proposed in literature is formulated by these steps:

$$\begin{aligned} \mathbf{v}_{n+1/2} &= \mathbf{v}_{n-1/2} + \Delta\tau \mathbf{a}_n \\ \mathbf{r}_{n+1} &= \mathbf{r}_n + \Delta\tau \mathbf{v}_{n+1/2} \end{aligned} \tag{3.50}$$

In this case the velocities have to be calculated as mean from the mid-step values

$$\mathbf{v}_n = \frac{1}{2}\{\mathbf{v}_{n-1/2} + \mathbf{v}_{n+1/2}\} \tag{3.51}$$

3.7.2.2 Nordsieck Algorithm

Nordsieck developed a specific algorithm for the solution of Newton's equation of motion [16]. Within this predictor-corrector scheme a Taylor expansion is used for the positions, velocities and accelerations and further derivatives. We obtain

$$\begin{aligned} \mathbf{r}(\tau + \Delta\tau) &= \mathbf{r}(\tau) + \mathbf{v}(\tau)\Delta\tau + \frac{1}{2}\Delta\tau^2 \mathbf{a}(\tau) + \frac{1}{6}\Delta\tau^3 \mathbf{q}_3(\tau) \\ &+ \frac{1}{24}\Delta\tau^4 \mathbf{q}_4(\tau) + \frac{1}{120}\Delta\tau^5 \mathbf{q}_5(\tau) + \cdots + \frac{1}{k!}\Delta\tau^k \mathbf{q}_k(\tau) + \cdots \end{aligned} \tag{3.52}$$

$$\begin{aligned} \mathbf{v}(\tau + \Delta\tau) &= \mathbf{v}(\tau) + \mathbf{a}(\tau)\Delta\tau + \frac{1}{2}\Delta\tau^2 \mathbf{q}_3(\tau) + \frac{1}{6}\Delta\tau^3 \mathbf{q}_4(\tau) \\ &+ \frac{1}{24}\Delta\tau^4 \mathbf{q}_5(\tau) + \frac{1}{120}\Delta\tau^5 \mathbf{q}_6(\tau) + \cdots + \frac{1}{(k-1)!}\Delta\tau^{k-1} \mathbf{q}_k(\tau) + \cdots \end{aligned} \tag{3.53}$$

$$\begin{aligned} \mathbf{a}(\tau + \Delta\tau) &= \mathbf{a}(\tau) + \mathbf{q}_3(\tau)\Delta\tau + \frac{1}{2}\Delta\tau^2 \mathbf{q}_4(\tau) + \frac{1}{6}\Delta\tau^3 \mathbf{q}_5(\tau) \\ &+ \cdots + \frac{1}{(k-2)!}\Delta\tau^{k-2} \mathbf{q}_k(\tau) + \cdots \end{aligned} \tag{3.54}$$

$$\begin{aligned} \mathbf{q}_i(\tau + \Delta\tau) &= \mathbf{q}_i(\tau) + \Delta\tau\, \mathbf{q}_{i+i} + \frac{1}{2}\Delta\tau^2 \mathbf{q}_{i+2}(\tau) \\ &+ \cdots + \frac{1}{(k-i)!}\Delta\tau^{k-i} \mathbf{q}_k(\tau) + \cdots \\ i &= 3.4, \cdots \end{aligned} \tag{3.55}$$

with

$$\mathbf{q}_k(\tau) = \frac{\partial^k}{\partial \tau^k}\mathbf{r}(\tau) \tag{3.56}$$

Using $\mathbf{r}^{(0)}$ and the scaled derivatives $\mathbf{r}^{(k)}$, which are defined by the simple expressions

$$\begin{aligned}
\mathbf{r}^{(0)} &= \mathbf{r} \\
\mathbf{r}^{(1)} &= \Delta\tau\mathbf{v} \\
\mathbf{r}^{(2)} &= 1/2\Delta\tau^2\,\mathbf{a} \\
\mathbf{r}^{(3)} &= 1/6\Delta\tau^3\mathbf{q}_3 \\
&\vdots \\
\mathbf{r}^{(k)} &= \tfrac{1}{k!}\Delta\tau^k\mathbf{q}_k,
\end{aligned} \tag{3.57}$$

we obtain a relatively simple Taylor series predictor that can be expressed through

$$\begin{Bmatrix} \tilde{\mathbf{r}}_{n+1}^{(0)} \\ \tilde{\mathbf{r}}_{n+1}^{(1)} \\ \tilde{\mathbf{r}}_{n+1}^{(2)} \\ \vdots \end{Bmatrix} = \mathbf{P}\begin{Bmatrix} \tilde{\mathbf{r}}_{n}^{(0)} \\ \tilde{\mathbf{r}}_{n}^{(1)} \\ \tilde{\mathbf{r}}_{n}^{(2)} \\ \vdots \end{Bmatrix} \tag{3.58}$$

$\mathbf{P}$ is the Pascal triangle matrix; it has binomial coefficients in its columns and is formulated by

$$\mathbf{P} = \begin{bmatrix} 1 & 1 & 1 & 1 & 1 & 1 & \cdots \\ 0 & 1 & 2 & 3 & 4 & 5 & \cdots \\ 0 & 0 & 1 & 3 & 6 & 10 & \cdots \\ 0 & 0 & 0 & 1 & 4 & 10 & \cdots \\ 0 & 0 & 0 & 0 & 1 & 5 & \cdots \\ 0 & 0 & 0 & 0 & 0 & 1 & \cdots \\ \vdots & \vdots & \vdots & \vdots & \vdots & \vdots & \ddots \end{bmatrix} \tag{3.59}$$

The predictor does not generate the exact values for the positions and their derivatives. However, using the predicted position $\tilde{\mathbf{r}}_{n+1}^{(0)}$ the forces for the time step

$n+1$ can be determined and, therefore, the correct values for the accelerations $\mathbf{a}_{n+1}$.

Corrector Steps:
The predicted results have to be corrected, and the corrector steps may be repeated in order to improve the results. How is the corrector step defined?

The comparison with the predicted accelerations $\tilde{\mathbf{r}}^{(2)}_{n+1}$ from Eq. (3.58) gives the possibility to estimate the error of the predictor step:

$$\vec{\varepsilon}_{n+1} = \frac{1}{2}\Delta\tau^2 \mathbf{a}_{n+1} - \tilde{\mathbf{r}}^{(2)}_{n+1} \tag{3.60}$$

The error $\vec{\varepsilon}_{n+1}$ can be used to improve the predicted values, and we get the following corrector step:

$$\begin{Bmatrix} \mathbf{r}^{(0)}_{n+1} \\ \mathbf{r}^{(1)}_{n+1} \\ \mathbf{r}^{(2)}_{n+1} \\ \vdots \end{Bmatrix} = \begin{Bmatrix} \tilde{\mathbf{r}}^{(0)}_{n+1} \\ \tilde{\mathbf{r}}^{(1)}_{n+1} \\ \tilde{\mathbf{r}}^{(2)}_{n+1} \\ \vdots \end{Bmatrix} + \begin{Bmatrix} c_0 \\ c_1 \\ c_2 \\ \vdots \end{Bmatrix} \vec{\varepsilon}_{n+1} \tag{3.61}$$

Within the predictor-corrector algorithm the corrector step may be repeated in order to improve the results. More details in connection with predictor-corrector methods are given in the [16] and the literature therein.

The predictor-corrector method for the treatment of differential equations are not restricted to molecular dynamics with respect to Newton's equations of motion, but has been applied so far for the solution of various other problems as, for example, for the treatment of Burger's equation.

3.7.3 Other Computational Methods

Molecular dynamics can be used for the investigation of classical systems and in fact most many-particle systems behave classically. Nevertheless, also "quantum molecular dynamics" is relevant. Furthermore, in the case of more complex systems "multiscale modeling" is an important method for the analysis of larger many-particle systems. Quantum molecular dynamics and also multiscale modeling will be briefly discussed in this section.

3.7.3.1 Molecular Dynamics with Quantum Effects

Car and Parinello developed a computational method, which directly combines quantum theory with classical mechanics. The equations of DFT (density functional formalism) are solved simultaneously to the classical equations of motion. This method is called quantum molecular dynamics. Here the classical Lagrange equation of the atomic positions and velocities are extended through the wave functions of Kohn and Sham and their time derivations. Such quantum molecular dynamical calculations are relatively complicated and, therefore, this method is restricted to a few hundred of particles. Nevertheless, the results are convincing.

3.7.3.2 Multiscale Modeling

How does nano-scale properties give rise to large scale behaviour? The answer can be found on the basis multiscale modelling. Here the properties over a wide range of length and time scales have to be studied. For this purpose certain strategies have been developed. For such multiscale modelling atomistic and coarse-grained simulations have to be done.

In order to be able to explain and predict the properties of specific materials at dimensions, ranging from nano-meter to meter, multiscale modelling has to be used. Three length scales are of particular interest: Atomic scale (nano-meter), microscale (micro-meter), and the mesoscale (ranging from millimeter to meter). Multiscale modelling connects these scales. Such investigations are computer studies, and big data sets are produced also here and need to be analysed carfully.

In connection with multiscale modelling interesting systems have been studied as, for example, the biopolymer translocation in connection with nanopores. This is a typical multiscale problem. On the one hand, the structure and dynamics of the polymer-pore system at the microscopic level is necessary to know. On the other hand, the processes on a larger scale have to be clarified; here molecular filtering, the protein transport within membranes are essential and we should know the mechanism that take place on a large time scale. For this set of problems coarse-grain models are important to develop, that is, through a bottom-up approach starting with atomistic configurations (produced by molecular dynamics). Also here big data sets come of course into play and data analysis is necessary in order to get a reduced form and observable information. In conclusion, multiscale modelling is an important investigation tool for the determination of properties that take place at various space and time scales. Polymer-pore systems are typical examples.

3.8 Final Remarks

Are big data sets unavoidable in many-particle physics? This is an interesting question and should be answered adequately. Big data sets in connection with many-particle systems come into play when the number of particles are large and are unavoidable due to the interaction between the particles of the many-particle system.

The following question arise: Is it possible to investigate such systems without the use of interacting particles? In this case we have to change the perspective. No doubt, this is a fundamental question and is relevant when we deal with big data sets.

The appearance of big data sets is dependent on the theoretical description. On the other hand, the theoretical description is dependent on the basic formulation of the physical problem, which is directly based on the facts that a human observer experiences. Is the present theoretical formulation of the physical laws the only possibility for human observers? This is certainly not the case; alternatives are not new and always possible. This point needs more detailed explanations. In this last section, we would like to give some details with respect to new developments in this field that are, as we remarked, also of relevance in connection with problems dealing with big data sets and their analysis.

3.8.1 Scientific Realism

Human beings normally assume that the real objects, which exist in the world are located in space and time; it is assumed that they are embedded there. This reflects the so-called container principle, and the container itself is defined through space and time. No doubt, most human observers assume as a matter of course that the impressions before them represent the real, material world itself. Those people who know that it is merely a “picture” of the outside world, which is in front of them believe that the geometrical structures in the picture are identical with the characteristics in the real world outside.

What kind of facts are described by theoretical physics, in particular by Newton’s mechanics? Is it the final truth ultimately described? This is in fact normally assumed, and this is the reason why we try to find a world formula, which is able to describe everything. Is physics able at all to formulate that what we call ”absolute truth”? This point needs a deeper analysis. Let us do that by means of the classical description introduced by Newton. We discussed this approach in almost all details in the last sections.

Is that what we describe by Newton’s equations of motion the “absolute truth”? Do these equations really describe the processes in “basic reality”? In other words, does nature also work in this way? It would be too naive to assume that. We have to distinguish between the “absolute truth” and an “observer-dependent truth”.

We will demonstrate that by means of an example. We talked about “big data sets”, but we have to consider them as human-specific and another world view would probably lead to other models and other sets of data. Before we talk about details, let us first give a simple example and let us investigate the motion of the earth in its way around the sun. This example is completely equivalent to the many-particle systems treated in the sections before, but this example is simpler because only two objects are involved.

3.8.2 The Naive Point of View

Earth and sun attract each other, and for this law Newton was able to formulate a mathematical model. On the basis of this law and initial conditions (the position of the earth and its velocity at a certain time (see Sect. 3.4.1), we can determine the movement of the earth around the sun for all times τ, and we get realistic results and not only principle statements. Exactly the same holds for the many-particle systems, which we studied in the sections before by molecular dynamics.

In this way we can determine the position of the earth in space in the year 2030 or where it was earlier as, for example, in the year 1850. For the calculation of these data, we only have to solve Newton's equation, which is a simple relation and is given by

$$m_E(d^2\mathbf{r}_\mathrm{E}/d\tau^2) = -Gm_Em_S(\mathbf{r}_E - \mathbf{r}_S)/|\mathbf{r}_E - \mathbf{r}_S|^3, \tag{3.62}$$

where $\mathbf{r}_\mathrm{E}$ is the position vector of the earth and $\mathbf{r}_\mathrm{S}$ is the position vector of the sun; m_E and m_S are the parameters for their mass; G in (3.62) is the gravitational constant. More elements are not needed for the calculation of orbits, no more elements appear in (3.62). Once again, in this case Newton's equation of motion enable to calculate the motion of an object (earth) through space under the presence of another object (sun), i.e., an interaction (expressed through a gravitational field) between earth and sun is effective for all times.

Let us discuss this process in somewhat more detail. The following question arise: Does Newton's equation describe that what is called “absolute truth”? Let us assume that this is actually the case, that is, we would like to assume that the absolute (basic) reality is reached with the development of the law (3.62).

Under this assumption, the masses m_E and m_S and also the gravitational forces between them have to be considered as really existing in basic reality. This imagination makes no problems at first. Then, the solutions of the differential equation (3.62) have to be considered as a real fact existing in the world outside as elements of the absolute truth, that is, they have to be considered as the deepest ground of reality. Nothing speaks against this view even when we cannot prove these statements directly. That what we observe directly are the paths. But do we really observe the absolute truth? This question must be answered. So far we stated that the masses m_E and m_S, the gravitational forces and also the paths may be considered as absolute fact and truth.

However, this view cannot be correct for the following reason: If all the elements of (3.62) reflect the absolute truth, then we have consequently to assume that the real object (in our example the earth) solves in its motion through space continuously the differential equations expressed by (3.62).

Such an imagination is however ridiculous and has to be ruled out; such a view appears to be absurd. For the solution of (3.62) we need a computer, but there is no computer in the world outside that could do this work. The theory does not contain a mechanism for solving the equations with lightning speed and everywhere in space on the basis of simultaneous events; such a mechanism should belong to the theory itself, but this is of course not the case.

The masses do not solve in their movement through space incessantly differential equations as the theory demands when we read the situation in this way, i.e., when we assume that Eq. (3.62) reflects the absolute truth.

As we already remarked, such a scientific realism has to be judged as absurd, which is also reflected in the following statement by Rescher [17]: *As Herschel ruminated long ago, particles moving in mutual gravitational interaction are, as we human investigators see it forever solving differential equations which, if written out in full, might circle the* earth. No doubt, this is an interesting remark.

In conclusion, this simple analysis underlines that Newton's physical construction does not affect that what we have called basic reality; Newton's ideas do not reflect the "absolute truth". However, Newton's basic way of thinking is used in all developments in natural science. Therefore, that what we said in connection with Newton's theory is also valid for all theoretical developments in physics, even for newer theoretical formulations of the physical phenomena.

The distinguished epistemologist Nicholas Rescher formulated this fact by the following statement [17]: *Scientific realism is the doctrine that science describes the real world: that the world actually is as science takes it to be, and that its furnishings are as science envisages them to be .... It is quite clear that it is not...* No doubt, Rescher's view is of fundamental relevance, in particular in connection with Newton's equations.

The paths of objects and the physical parameters together with the phenomena tied with them, which are all formulated through Eq. (3.62) can only reflect a specific kind of truth, but this truth cannot reflect an absolute statement. We have to accept that and we have obviously no choice. The consequences are however radical: All the things, which directly appear before us (before our eyes) cannot be the last, absolute truth. This is, in other words, not basic reality itself and is of course also the case for the facts behind (3.62), i.e., the movements of the celestial bodies, which exactly appear before us in the form described by Eq. (3.62), may not be seen as an absolute fact; they do not reflect the last, deepest ground of reality. Then, also all the other phenomena before us, which we feel in particular by our senses, cannot be absolute facts, also that what we perceive of human beings and other creatures. These facts are basic in character.

In conclusion, the physical laws of motion cannot be the last truth. Human beings are caught in space and time and this means the following: Since the derivation of these laws have directly been based on our impressions in everyday life that we have directly in front of us, also these optical phenomena in everyday life cannot be part of an absolute, basic reality. Thus, a human being cannot recognize the absolute truth in this way; the deepest ground of the world outside is in principle not recognizable for a human being [18].

What kind of reality is behind Eq. (3.62)? Is it actually a specific kind of reality that we have directly before our eyes or is it "only" a symbolic picture of it? These and similar questions need particular attention and analysis.

The usual basis for the physical description of the world is to assume that everything is embedded in space and time. Then, the assumption that the phenomena before us reflect a real, absolute situation is by all means reasonable. However, it turned out that this should not be realistic (see in particular [18]). The analysis showed convincingly that the direct optical iphenomena before us, positioned in space and time, are "pictures of reality" but not basic reality itself, i.e., the objects in space and time are not directly the real masses but reflect geometrical figures.

3.8.3 The Projection-Principle

This "naive view of view" is not the language of science. The belief that this situation reflects the absolute truth (absolute reality) is due to the assumption that the container model is valid. "Container model" means that there is nothing outside space and time, that is, all what is inside the container can be judged as the "absolute truth". Here the real objects with certain masses are positioned in space and time, and the objects move relative to each other under the influence of interactions. This scenario, described by Newton's equation of motion, has to be considered as absolute truth when we work within the frame of the container conception. Since there can be nothing outside this container, we obtain that what is often called "basic reality", This statement is quite independent of the theoretical description, i.e., Newton's equations of motion have to be considered as the final physical description scheme, but only within the container model.

Quantum phenomena and other approaches are usually also represented within the framework of the container model and should also not be classified as the absolute truth.

However, we just recognized that the scientific realism behind this view suggest that this scenario is too naïve and should be rejected. The physical processes do not take place "within space and time" but are "projected on space and time". Not the container-principle (container model) but the more realistic projection-principle should be applied [18]. Convincing arguments are in favour of the projection-principle. Within this conception the "naïve point of view" can be eliminated. We don't want to discuss here the details, but only those items that are relevant in connection with many-particle systems and big data sets.

How can we characterize the main features of the projection-principle? For the projection-principle the following points and characteristics are essential:

1. The optical impressions in front of us (before our eyes), observed in everyday life, are pictures and is not reality itself. They are images of basic reality.
2. Basic reality exists but remains hidden to a human observer, and this due to the principles of evolution.
3. Basic reality can be replaced by a "model reality", which has to be constructed by the human observer.

All human beings are caught in space and time and, therefore, the construction of the "model reality" has to be based on space and time. A detailed analysis showed [18] that the variables of the model reality are defined by the momentum and the energy:

$$\mathbf{p} = (p_x, p_y, p_z),\ E \tag{3.63}$$

These variables define the so-called (**p**, E)-space. Let us denote the information, embedded in this space, by $\psi(\mathbf{p}, E)$, which has the features of a wave function [18]. On the other hand, we have a "picture of reality", and this picture is represented in space and time:

$$\mathbf{r} = (x, y, z,),\ t \tag{3.64}$$

These variables define the so-called (**r**, t)-space. The information, located in this space, is denoted by $\psi(\mathbf{r}, t)$.

The relevant statement of projection theory can be formulated as follows: The information $\psi(\mathbf{p}, E)$ represented in (**p**, E)-space is projected onto (**r**, t)-space and gives $\psi(\mathbf{r}, t)$:

$$\psi(p_x, p_y, p_z, E) \to \psi(x, y, z, t) \tag{3.65}$$

The transition from (**p**, E)-space to (**r**, t)-space is given within the framework of projection theory by a Fourier transform:

$$\begin{aligned}\psi(x, y, z, t) = \frac{1}{(2\pi\hbar)^2} \int\limits_{-\infty}^{\infty} \psi(p_x, p_y, p_z, E) \\ \times \exp\left\{\frac{i}{\hbar}\left[(p_x x + p_y y + p_z z) - Et\right]\right\} dp_x dp_y dp_z dE\end{aligned} \tag{3.66}$$

The time t reflects a quantum-aspect of time, which is a new characteristic in the description of quantum systems and is a result of projection-theory. The variable t is different from the time τ we have used in connection with molecular dynamics; in contrast to t the time τ is a classical parameter and reflects a reference time in projection theory. However, we do not give more details here.

Analogy to General Theory of Relativity:
In the case of General Theory of Relativity the situation is similar and is comparable to that what we have said in connection with projection theory. This can be discussed in an easy way by means of Einstein's famous field equations, which have the following well-known form:

$$R_{\mu\nu} - \frac{1}{2} g_{\mu\nu} R = -\kappa T_{\mu\nu} \tag{3.67}$$

with

$$\kappa = 8\pi G \tag{3.68}$$

where G is Newton's gravitational constant. The other elements are defined as follows:

$g_{\mu\nu}$ Riemann's metric tensor; it is responsible for the geometry.
$T_{\mu\nu}$ momentum-energy tensor.
$R_{\mu\nu}$ Ricci tensor.
R scale factor, which determines the size of the universe.

The determination of the elements $g_{\mu\nu}$ can be done by Eq. (3.67); $g_{\mu\nu}$ is given as a function of the momentum-energy distribution. Then, reality [(($\mathbf{p}, E$)-distribution] determines the features of space, i.e., its geometry which is curved (non-Euclidian) within General Theory of Relativity.

Exactly the same scheme is valid in projection theory. The quantum phenomena in projection theory are formulated within $(\mathbf{p}, E)$-space ("reality") and the projection of this information on $(\mathbf{r}, t)$-space leads to the "picture of realty", however within an Euclidean geometry.

A specific $\mathbf{p}, E$-distribution in General Theory of Relativity leads to a curvature of space-time; each kind of curvature reflects a certain $\mathbf{p}, E$-distribution. Thus, reality can be interpreted as $\mathbf{p}, E$-distribution producing a curved space-time and can therefore be considered as the "picture of reality"; it reflects the $\mathbf{p}, E$-distribution through a specific curved space.

Then, we have the following connections between both theories, i.e., between General Theory of Relativity and projection theory.

General Theory of Relativity:

$$T_{\mu\nu}(reality) \rightarrow g_{\mu\nu}(picture\ of\ reality) \tag{3.69}$$

Projection Theory:

$$\begin{aligned} &\psi(p_x, p_y, p_z, E)\ (reality) \\ &\quad \rightarrow \psi(x, y, z, t)\ (picture\ of\ reality) \end{aligned} \tag{3.70}$$

That is, instead to assume that reality is positioned within space-time, we use also the projection-principle within General Theory of Relativity, but it is a re-interpretation of the original framework. All that is fundamental and needs further analysis.

Determination of the Wave Functions:
The functions $\psi(\mathbf{r}, t)$ and $\psi(\mathbf{p}, E)$ are of basic relevance within projection theory. If we know the interaction $V(\mathbf{r},t)$ between the system under investigation with external units, the functions $\psi(\mathbf{p}, E) = \psi(p_x, p_y, p_z, E)$ and $\psi(\mathbf{r}, t) = \psi(x, y, z, t)$ can be determined, at least in principle. These equations can be deduced within projection theory and take the following form [18]:

$$i\hbar\frac{\partial}{\partial t}\Psi(x,y,z,t) = -\frac{\hbar^2}{2m_0}\Delta\Psi(x,y,z,t) + V(x,y,z,t)\Psi(x,y,z,t) \tag{3.71}$$

and

$$\begin{aligned} E\,\Psi(p_x,p_y,p_z,E) &= \frac{\mathbf{p}^2}{2m_0}\Psi(p_x,p_y,p_z,E) \\ &+ V\left(i\hbar\frac{\partial}{\partial p_x}, i\hbar\frac{\partial}{\partial p_y}, i\hbar\frac{\partial}{\partial p_z}, -i\hbar\frac{\partial}{\partial E}\right)\Psi(p_x,p_y,p_z,E) \end{aligned} \tag{3.72}$$

where m_0 is the mass of the system. Both relations are equivalent and reflect exactly the same information. These equations have been derived in [18], but we will not discuss the details here. Let us only mention that the expression

$$-i\hbar\frac{\partial}{\partial E}$$

in Eq. (3.72) is the operator for the time-coordinate that is an outcome of projection theory.

The Unified Whole:
When we work within the container conception the world seems to exist in the form of separated objects. This is in fact our impression that we have in everyday life. Newton exactly used this conception and his equations of motion reflect that: We have N objects that interact with each other. Within molecular dynamics, which we discussed in the sections before, Newton's conception has been used, and the solution of these equations leads in all cases to "big data sets".

When we work within projection theory we cannot assume that the real objects exist in basic reality as our impressions suggest and dictate when we observe the world in space and time. Although we know only a little about basic reality, we may assume that its structure is not granular in character; granular structures are space-time features and should not be transferred necessarily to basic reality. The system under investigation interacts in basic reality with an external system, and the projection onto space and time results in a wave function $\psi(\mathbf{r}, t)$ that show certain structures. If these structures have definite peaks at certain space-time positions, say

$$\mathbf{r}_1, t_1, \mathbf{r}_2, t_2, \mathbf{r}_3, t_3, \ldots, \mathbf{r}_N, t_N, \tag{3.73}$$

we may approximate $\psi(\mathbf{r}, t)$ by a function

$$\Phi(\mathbf{r}_1, t_1, \mathbf{r}_2, t_2, \mathbf{r}_3, t_3, \ldots, \mathbf{r}_N, t_N), \tag{3.74}$$

which consists for example of N particles that interact with each other. In other words, we have the following transition:

$$\psi(\mathbf{r}, t) \Rightarrow \Phi(\mathbf{r}_1, t_1, \mathbf{r}_2, t_2, \mathbf{r}_3, t_3, \ldots, \mathbf{r}_N, t_N) \tag{3.75}$$

In this case we may introduce equations of motions in analogy to Newton's basic equations and we get for each time τ a set of $\mathbf{r}, t$-positions. (Note that the system-specific time t is not defined in traditional physics, but merely the reference time τ that we measure with our clocks.) In this way, we get "big data sets". However, we know that the source for this approach is not (3.74), but the wave function $\psi(\mathbf{r}, t)$ that is obviously much simpler than (3.74), and the use of $\psi(\mathbf{r}, t)$ instead of (3.74) can lead to a strongly reduced data set, although exactly the same physical problem is analyzed.

3.9 Summary

In solid state physics, material science, biology, nanoscience and nanotechnology we deal with systems. For the theoretical analysis of such systems the solution of Newton's theory is relevant. This can be done in the most general form by means of molecular dynamics using suitable predictor-corrector algorithms. In this way "big data sets" are produced. This large amount of information has to be analyzed and reduced.

Statistical-mechanical ensembles have been introduced in order to get a relationship between the big data set and physical properties, in particular in connection with measureable quantities. Big data sets are possibly then avoidable if we change the frame of description. It is obvious to replace the container-principle through the projection-principle. A brief description is given, and the relevance for big data sets are underlined.

Data analysis is relevant when we deal with big data sets in connection with many-particle systems. Concerning data analysis etc., excellent scientific work has been done and published [19–22], but could not be discussed in this chapter.

References

1. Rieth M, Schommers W (2006) Atomic nanodesign, handbook of theoretical and computational nanotechnology. American Scientific Publishers, Valencia
2. Torrens M (1972) Interatomic potentials. Academic Press, New York
3. Daw MS, Hatcher R (1985) Application of the embedded atom method to phonons in transition metals. Solid State Commun 56:697–699
4. Foiles SM (1985) Application of the embedded-atom method to liquid transition metal. Phys Rev B 32:3409–3415
5. Daw MS (1986) Surface reconstruction and many-atom models. Surf Sci 166:L161–L164
6. Arregui EO, Caro M, Caro A (2002) Characterizing many-body localization by out-of-time-ordered correlation. Phys Rev B 6:054201
7. Schommers W (1975) The volocity autocorrelation function and the effect on the long-range interaction in liquid rubidium. Solid State Commun 16:45–47
8. Lüscher E (1973) in Die feste Materie, Umschau Verlag, Frankfurt am Main
9. Morse PM (1929) Diatomic molecules according to the wave mechanics. Phys Rev 34:57–60
10. Mohammed KMM, Shukla F, Milstein J, Merz L (1984) Resonant phonon-assisted energy transfer in ruby from 29-cm^{-1}-phonon dynamics. Phys Rev 29:3117–3122
11. Eichler M, Peyzl M (1969) Intrinsic stacking faults on 112 planes in the B.C.C. Lattice Phys Stat. Sol 35:333–338
12. Lincoln RG, Koliwad KM, Ghate PB (1967) Morse-potential evaluation of elastic constants of some cubic metals. Phys Rev. 157:463–466
13. Schommers W (1999) Excited nano-clusters. Appl. Phys. A 68:187–196
14. Schommers W (1976) The Effect of van der Waals-Type interactions in metals: a pseudopotential Model. Z. Phys. B 24:171
15. Schommers W (1997) Phonons and structure in nano-clusters: a molecular dynamics study for Al. Nanostruct. Materials 9:693–696
16. Rieth M (2003) In: Politis C, Schommers W (eds) Nano-engineering in science and technology, series on the foundations of natural science and technology, vol 5. World Scientific, New Jersey, London
17. Rescher N (1977) The limits of science. University of California Press, Berkeley
18. Schommers W (2011) Quantum processes. Word Scientific, London
19. Debi A, Anitha A (2017) A comparative study of statistical and rough computing models in predictive data analysis. Int J Ambient Comput Intell 8:32–51
20. Kamal S, Ripon SH, Dey N, Ashour AS, Santhi V (2016) Comput Mehods Programs Biomed 131:32
21. Zhang W, Qi Q, Deng J (2017) Building intelligent transportation cloud data center based on SOA. Int J Ambient Comput Intell 8:1–11
22. Dey N, Hassanie AE, Bhatt C, Ashour A, Satapathy SC (eds) (2017) Internet of Things and big data analytics toward next-generation intelligence. Springer, Berlin

Chapter 4
Earth Science [Big] Data Analytics

Mani Madhukar and Pooja

Abstract Tremendous research has been done and is still in progress in the domain of earth science. With the advent of Big Data and availability of datasets on Earth science, the study of Earth sciences has reached new dimensions. The diversity and high dimensional remote sensing data have provided with complex data sets capable of giving insights and intelligence that was not possible in last decades. With Computing progress made in ingesting and inferring data from myriad of sources including high resolution cameras mounted on satellites and sensors giving access to unconventional Big Data and also with the GPU computing and Data science advances we are today able to leverage machine learning and deep learning in extensively complex datasets derived from remote sensing about Earth Sciences. Our focus is to analyze what exactly does big data mean in earth science applications and how can big data provide added value in this context. Furthermore, this chapter demonstrates various big data tools which can be mapped with various techniques to be used for experimenting earth science datasets, processed, and exploited for different earth science applications. In order to illustrate the aforementioned aspects, instances are presented in order to demonstrate the use of big data in remote sensing. Firstly, this chapter presents earth science studies, application areas/research fields and a brief insight on earth science data. Then various techniques implemented in this domain are elaborated viz. classification, clustering, regression, deep learning, pattern recognition, machine learning, earth data analysis and processing. Later this chapter introduces big data analytics and various tools/platforms in big data viz. BigInsights, GIS, Jupyter notebook, Matlab, Python. Finally, it is shown how these tools are mapped to Earth science datasets using ArcGIS to illustrate with experimental instances the inferences and patterns generated.

M. Madhukar (✉)
IBM India Pvt. Ltd., Bangalore, India
e-mail: manimmadhukar@gmail.com

Pooja
NIT Jalandhar, Jalandhar, India
e-mail: poojachoudhary80@gmail.com

N. Dey et al. (eds.), *Big Data for Remote Sensing: Visualization, Analysis and Interpretation*, https://doi.org/10.1007/978-3-319-89923-7_4

Keywords Earth data analysis and processing · Geosciences · GIS
Big data analytics · Python · MATLAB · Jupyter notebook · BigInsights
ArcGIS

4.1 Introduction

The study of earth sciences or geosciences deals with the various fields of science related to our planet Earth in terms of its physical constitution and the atmosphere surrounding it. Typically, Earth science is a division of the physical sciences (a branch of natural sciences) which only deals with the study of non-living systems (Fig. 4.1).

The official discipline of Earth sciences embraces the study of the biosphere, hydrosphere, atmosphere, oceans as well as the concrete earth. Characteristically, Earth sciences is the quantitative understanding of the working of the Earth system,

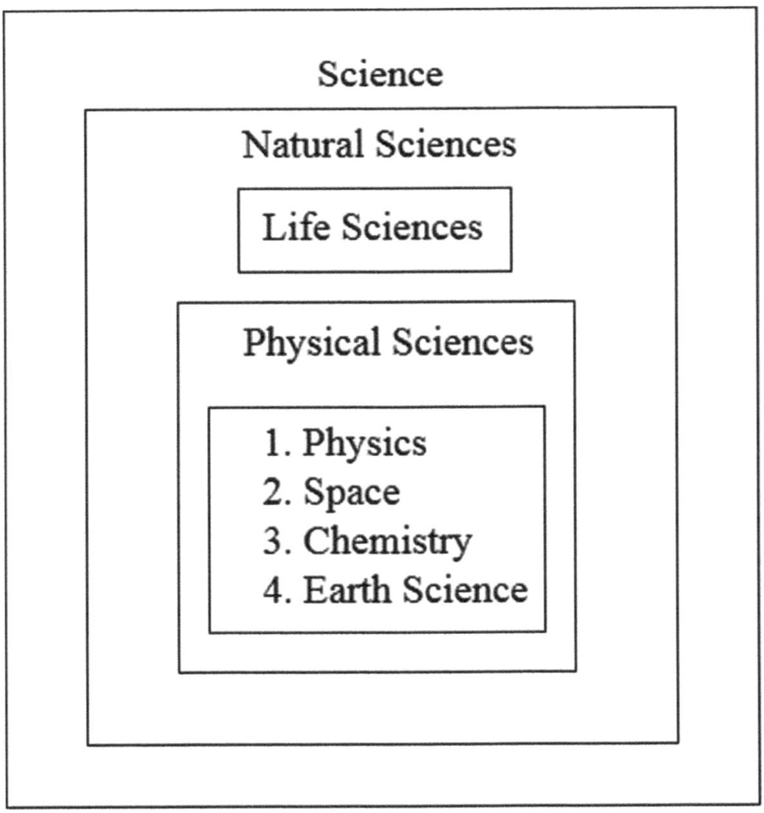

Fig. 4.1 Hierarchical positioning of earth science as a branch of science

and development of earth system to its current state. Earth Science studies includes (but not limited to):

- Identifying climate change and its remarkable consequences.
- Examining geo-hazards and forecasting earthquakes and volcanic eruptions.
- Finding out natural resources like hydrocarbons and valued mineral deposits and managing these resources.
- Managing groundwater and other natural water supplies.
- Understanding Pollution levels (air, water, soil) and its impact on ecosystem.
- Environmental sustainability.
- Exploring the strength of interior earth like bedrock.

4.2 Field of Studies of Earth Sciences

The Earth sciences can be catagorized into following:

- The study of soil, study of water (hydrology), study of climate (climatology), geomorphology, meteorology, and biogeography comes under Physical geography.
- Hydrology includes study of both freshwater and oceanic waters (limnology and oceanography).
- Study of lithosphere comes under Geology which has subdisciplines like mineralog, structural geology, sedimentology to name a few.
- Geophysics: study os earth's core and mantle.
- Geodesy is exploring the Earth's shape (including magnetic and gravity fields), earth's reaction to external forces.
- Study of pedosphere (outer layer of crust) is Soil science.
- Ecology covering the connection of biota and natural environment.
- Glaciology covers cryosphere.
- Atmospheric-sciences encompasses the atmosphere and it includes atmospheric physics, atmospheric chemistry, meteorology and climatology.

Earth science includes the study of all that comes under the umberella of four spheres of earth viz. the atmosphere, the biosphere, the hydrosphere and the geosphere and may include cryosphere (the study of ice). Earth science in itself is a huge domain to study about and encompassing, the deteriorating changes in the Earth and its atmosphere, endangering the life on Earth, which has raised the alram to find ways to fight against them for the sustainability and betterment of human race. Understanding the processes occurrence in the interior of the Earth, on the earth surface and in its surrounding atmosphere is complex but considerable. This knowledge about the composition of the Earth, and their linkages with geophysical observations and its dynamic state has highly advanced in the most recent decades, and has produced huge data over the years about the weather conditions, climate, floods, hurricanes, increasing carbon rate emissions, movement in tectonic plates, glaciers etc. Studying this newly discovered data can surely lead us to better methods and techniques to avoid hazards and dangers posed to planet Earth in future.

4.3 Earth Science Data

Earth scientists gather a range of information related to the surface of Earth, make observations, collect data about natural processes on Earth. This data is obtained by means of different acquisition methods (space and time domain), having varying data features. Also, future changes of planet are predicted based on analysis done on collected. Collecting Earth Sciences data has a long record going back to centuries. Initially making pen-paper records to Scientific instruments based records to imagery data records from satellites. And then high-resolution imagery records for good quality and better results have lead to drastically increased amount and size of imagery data (may term it as ‘Earth Science Big Data’).

Since this Earth science data has huge volume and variety, it is essential to recognize their type to effectively use of bigdata analytics approaches for research in this domain.

‘Earth science data can be broadly categorized into following three categories viz. observational data, climate simulations and reanalysis data’ [1].

Observational datasets are the outcome of the sensor measurements of sources like satellites or ground based stations. Mostly such datasets can be grouped as gridded and station-based data.

Station based data:

- error prone
- direct data
- obtained over irregularly and differently placed stations like weather station.

Gridded data:

- postprocessed using various techniques like aggregation, interpolation and sampling techniques
- easily accessible

Climate simulations are produced using substantial the Earth’s climatic process models.

Reanalysis data involve analyzing the irregularly recorded observations and interpolating the missing value (poor quality observations/large area values) by creating physical models along with.

Processing the data so gathered is challenging as it is deteriorated due to noise, sensor interference and may be due to system malfunction. Furthermore, Earth science datasets are captured at multiple resolutions and multiple scales as Earth's surface appears to change at different spatial and temporal scales.

4.4 Earth Science Data Analytics, Tools, and Techniques

Big Data Analytics is about the capability to study bulk of information of a range of types to explore unknown pattern, unidentified correlations, and other constructive information.

'The process of examining, preparing, reducing, and analyzing large amounts of spatial (multi-dimensional), temporal, or spectral data using a variety of data types to uncover patterns, correlations, and other information, to better understand our Earth.

Data Preparation—Preparing heterogeneous data so that they can be jointly analyzed.

Data Reduction—Correcting, Ordering, and Simplifying data in support of analytic objectives.

Data Analysis—Applying techniques/methods to derive results' (Table 4.1) [2].

4.4.1 Various Approaches for Earth Data Analytics

Machine Learning: (ML) it falls under computer science domain and this subfield evolved from pattern recognition and computational learning theory. Machine algorithms are based on learning that can be supervised or unsupervised learning. Based on the learning, ML algorithms make predictions on data. Machine learning itself is a very vast and rapidly growing field. Machine Learning algorithms derive intelligence from data directly, and thus helps in making data-driven decisions. Machine learning based techniques are advantageous than conventional statistical methods as these do

Table 4.1 Analytics tools and techniques

Analytics	Data preparation	Data reduction	Data analysis
Tools	R, SAS, Python, Java, C++, SPSS, MATLAB, Minitab	VBA, Excel, MySQL	Hadoop, GIS, Parallel NetCDF, Javascript, Perl, PHP, Cloud solutions, AWS, open source databases
Techniques	Statistics functions, machine learning, factor analysis, principal component analysis, natural language processing, neural networks	Linear/ non-linear regression, text analytics	Logical regression, time-series analysis, decision tree, clustering, MapReduce, graph analytics

not impose impractical suppositions like linearity. Further ML algorithms can infer omitted data, thus reducing the overhead of long-term expert annotation.

Statistical modeling and techniques: Statistical modeling can be considered as formal mathematical way to approximate reality. Statistical modeling explains the generation of data and extends to making predictions from these approximations. In other terms it is a simplified, mathematically formulized way to create statistical models from mathematical equations using explanatory and dependent variables.

- In general, Regression is a statistical measure which is used in difference disciplines to determine the strength of the relationship between one dependent variable and a series of other changing variables, which are independent variables. Regression analysis helps to explain how the value of the dependent variable changes when any one of the independent variables is varied, while the other independent variables are held fixed. **Regression analysis** is a part of statistical modeling defining set of statistical processes for estimating the relationships among variables. It includes many techniques for modeling and analyzing several variables, when the focus is on the relationship between a dependent variable and one or more independent variables. In all cases, a function of the independent variables called the **regression function** is to be estimated. In regression analysis, it is also of interest to characterize the variation of the dependent variable around the prediction of the regression function using a probability distribution.

Regression analysis is primarily used for prediction and forecasting based on learning from historical data. The use of Regression analysis has substantial overlap with machine learning also. Regression analysis is used to understand which among the independent variables are related to the dependent variable, and to explore the forms of these relationships.

In restricted circumstances, regression analysis can be used to infer causal relationships between the independent and dependent variables. However, this can lead to illusions or false relationships, so caution is advisable, for example, correlation does not imply causation. There has been a lot of study on use of Regression in Earth Science, and it is cautioned against getting misleading results in some cases. Common methods of Least square and Linear regression are used for Earth Science data analytics beside others.

- **Decision Trees**: are graphs that use branching method to illustrate every possible outcome of a decision. It is decision support tool which uses tree like graph to show possible outcomes. Decision trees are commonly used in Operations research for decision analysis and identifying suitable strategy to achieve a certain goal. A decision tree consists of three types of nodes:

  1. Squares representing 'Decision nodes'.
  2. Circles representing 'Chance nodes'.
  3. Triangles representing 'End nodes'.

Decision trees can be drawn by hand or produced using some graphics program or software. They are used to simplify complex strategic challenges, evaluate the cost effectiveness of decisions (research and business). In machine learning also they are useful.

- **Time series analysis**: As mentioned in the introduction section, Earth science data is collected in space as well as time domain, as it adheres to the spatial and temporal changes. Time series analysis is another statistical technique to analyse data which is captured over a period or at regular time intervals. We normally cover three types of data for time series analysis:

  1. **Time series data**: time series data is a group of values for particular variable recorded at diverse time. A time series plot simply plots some variables against time. For example, record of earthquakes occurred over past twenty years.
  2. **Cross-sectional data**: is the record of one (or more) variables gathered at one point of time.
  3. **Pooled data**: is a blend of time series and cross-sectional data.

Time series analysis is also referred to as trend analysis. Figure 4.2 is a **time series plot** of the world earthquakes with magnitude 7.0–7.9, from year 1970–2012. Data collected from [3].

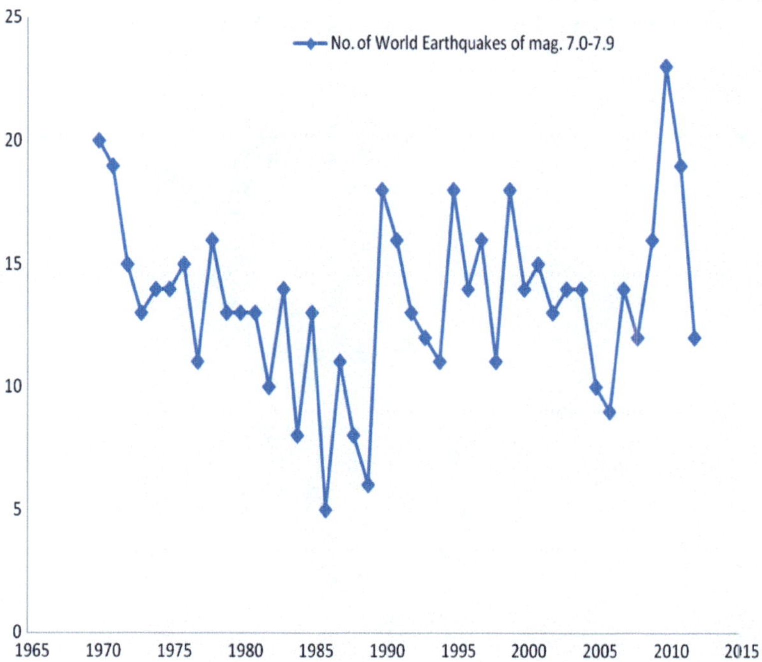

Fig. 4.2 Time series plot of World earthquakes with magnitude over 7.0–7.9 from 1970 to 2012

Data mining techniques: The quantity of raw data being stored today by different archives of Earth-Science, defies even the partial manual examination by scientists. Data driven techniques have found their way in many Earth sciences subdivisions including hydrology, hydrogeology, natural disasters, geomorphology, meteorology, soil study.

There are two major data mining tasks first is prediction methods i.e. predicting future value using some known variables and second is description methods i.e. finding patterns that describe the data. Various data mining techniques are:

- **Classification** allocates item in a set to target categories or classes. The purpose of classification is to predict the target class for every case in the data accurately. The problem of recognizing a set of categories (sub-populations) for a given item/test data, to which it belongs. Classification is one of the most widely used techniques. It can be of two types; two class classification (when prediction is made out from two categories) and multi-class classification (when there are more than two categories). It is a supervised learning type of machine learning. Classification can be used, for example to classify, land covers, water bodies, urban areas, forests, etc. using remote sensing data. 'Classification maps were generated for three years [4] authors have used ENVI software to carry out image classification. A supervised classification technique with Maximum Likelihood Algorithm was applied'. Figure 4.3 displays land-use/cover classification map for year 2010, for middle India areas using ArcGIS.

There are many classification techniques available viz. k-nearest neighbor, Support vector Machines, Linear classifiers, Neural Networks, Decision Trees.

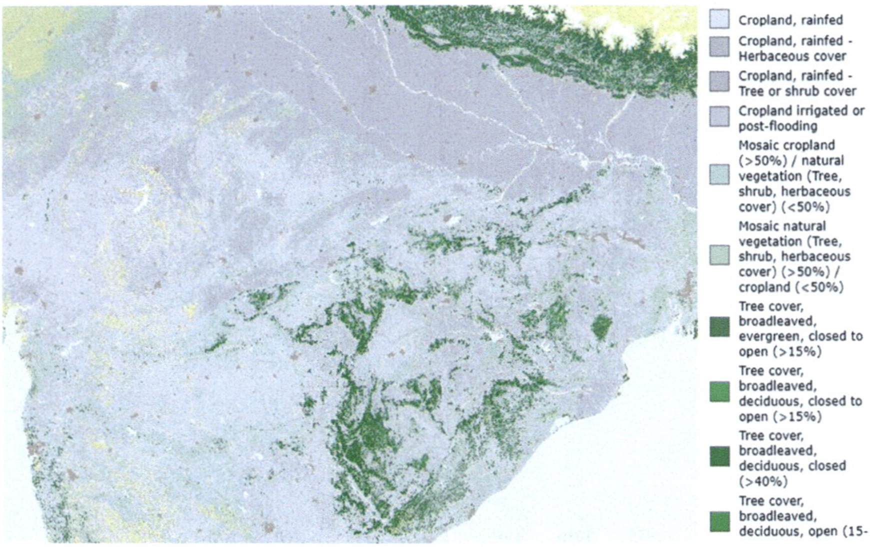

Fig. 4.3 Land-use/cover classification maps for Middle India (2010 year)

- **Clustering**: A cluster is a collection of objects that belong to same class. It is an assembly of group of objects (of same features) combined together based on similarity. When implemented for data mining, clustering categorizes data by applying some algorithm which is suitable for required analysis. In Earth Science data mining is an essential task to find interesting patterns in Earth science data. These patterns are used for better understanding and predicting changes for example in global water cycle, carbon emission cycle or weather conditions. In this regard clustering is used to divide the Earth geographical areas (land and water) into different regions for discovering interesting patterns directly or indirectly. For example, in the paper titled 'The Application of Clustering to Earth Science Data: Progress and Challenges', authors have created 'twenty-five number of ocean clusters of sea level pressure data for 1958–1998, as produced by shared nearest neighbor clustering' [5].

'Clustering is divided into hierarchical clustering and non-hierarchical clustering. In hierarchical clustering, the similarity of a cluster is evaluated using a "distance" measure' [6]. Nearest neighbor method is an example of hierarchical clustering. While in non-hierarchical clustering initially, an arbitrary number of clusters should be temporally chosen. 'The members belonging to each cluster are checked by selected parameters or distance and relocated into the more appropriate clusters with higher separation. K-mean method is an example of non-hierarchical clustering' [6].

Figure 4.4 displays Self organizing map (SOM) based clustering performed on air pollution mortality dataset.

- **Evolutionary Algorithms**: 'An Evolutionary Algorithm (EA) is essentially an algorithm inspired by the principle of natural selection and natural genetics. The basic idea is simple. In nature individuals are continuously evolving, getting increasingly adapted to the environment' [7]. 'The main role of evolutionary algorithms in most of these approaches is optimization. They are used to improve the robustness and accuracy of some of the traditional data mining techniques. Different types of evolutionary algorithms have been developed over the years such as genetic algorithms, genetic programming, evolution strategies, evolutionary programming, evolution strategies, differential evolution, cultural evolution algorithms and co-evolutionary algorithms' [8] 'Some of these types that are used in data mining are genetic algorithms, genetic programming, and co-evolutionary algorithms. Genetic algorithms are used for data preprocessing and for post processing the discovered knowledge, while genetic programming is used for rule discovery and data preprocessing' [9].

For example, in their chapter titled "The Application of Fuzzy Logic and Genetic Algorithms to Reservoir Characterization and Modeling" [10], authors have mentioned that "shear sonic data is required for rock strength analysis to determine fracture propagation and formation breakdown characteristics, and for improved porosity prediction as shear sonic data is largely unaffected by fluid type. Shear

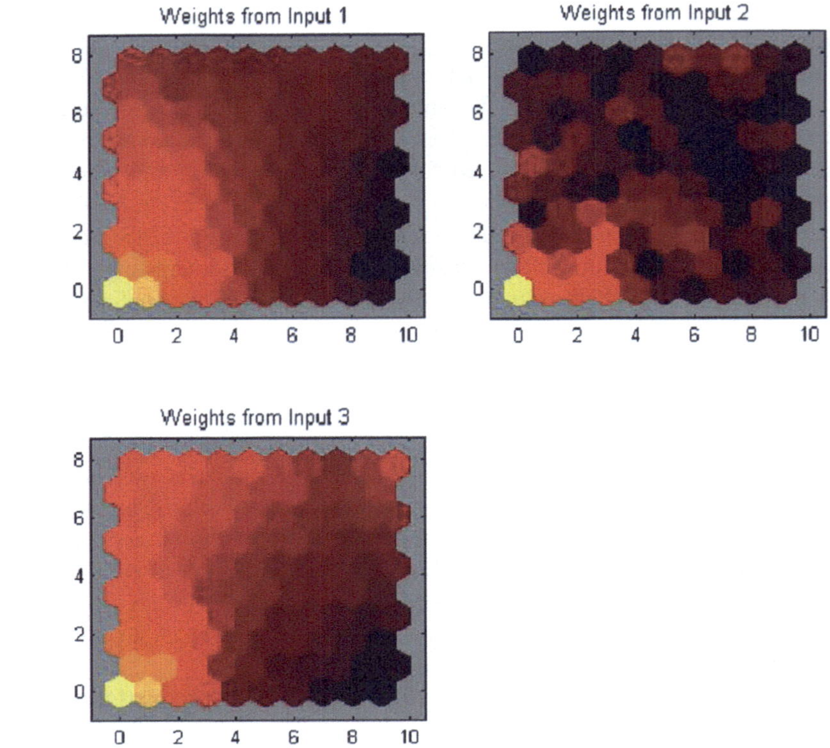

1

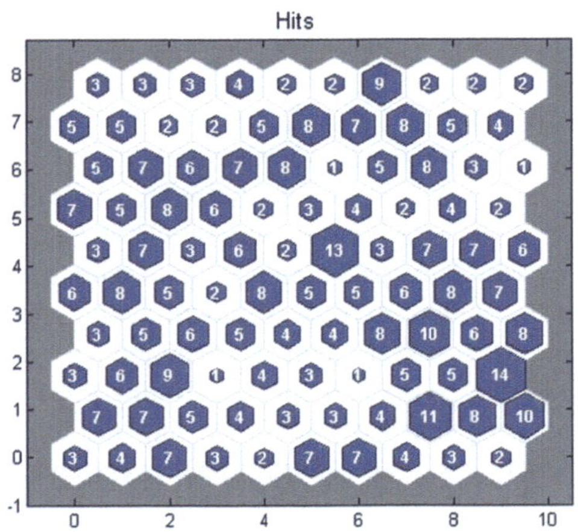

2

Fig. 4.4 SOM clustering on air pollution morality dataset, (1) weight plane of element of the input vector (three). Darker colors represent larger weights and (2) 100 neuron locations (clusters) and data associated with each cluster centre. The topology is a 10-by-10 grid, so there are 100 neurons sonic data are also becoming important for enhanced seismic interpretation. Because the value of shear velocity data is only now being realized, and because such data is expensive to acquire, there is limited amount of information available in the North Sea. Genetic algorithms have been used to determine the shear velocities in oil wells based on calibrations elsewhere in the oil field. Not only have genetic algorithms determined the constant parameters of these calibrations, but GAs have also evolved the calibration equations themselves".

Pattern Recognition: is a branch of machine learning and involves identification of patterns of data. A pattern is characteristics of the data such as shape, texture, regularity etc. Pattern recognition generally includes classifying the patterns from the data, segmentation of the pattern out of the overall data and the finally identifying these patterns using labels or identifiers. In some cases, pattern recognition and machine learning are used as synonyms. Statistical pattern recognition is gaining popularity in modern computer science. It deals with exploring, analysing, and recognizing patterns from complex data like high resolution images (from satellites for remote sensing) etc.

Deep Learning: It is a type of machine learning only and deep learning based system tends to perform classification directly from the provided data (images, text, sound) under observation. Deep Learning is based on neural network architecture, in deep learning there are enormous number of hidden layer unlike 2–3 in conventional neural network. Each layer of deep learning network receives data from the preceding layer, transforms it, and passes it to the next following layer. In this way, the network learns directly from the data at every layer which in turns increases the complexity and detail of what it is learning from layer to layer, therefore making the system more efficient and intelligent.

4.4.2 Categories of Analytics

Data analytics can be broadly categorised in three major divisions, as represented in Fig. 4.5, viz. Descriptive, Predictive, and Prescriptive analytics.

- **Descriptive analytics** focuses on preparing data for advance analytics and it includes varieties of methods including regression, data modelling and visualization.
- **Predictive analytics** involves predictive modelling and data mining. It focuses on deriving predictions using statistical techniques such as linear or logistic regression, extracting data patterns using data mining techniques.
- **Prescriptive analytics** primarily includes decision making and focuses on efficiency. It makes use of optimization and simulation techniques to achieve efficiency.

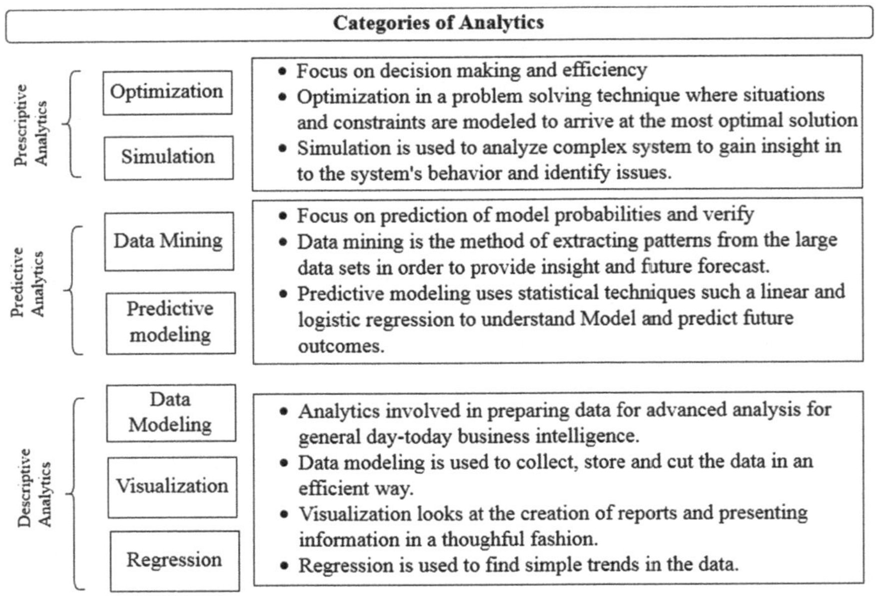

Fig. 4.5 Categories of analytics

4.5 Some Powerful Tools for Analytics in Earth Sciences

In this section, we will elaborate on few tools which are used in Earth Science data analytics. The adoption of these tools has led to widespread insights being obtained by global leader working on Earth Science analysis in Oil and Gas exploration, studying changes in different domains of Earth Science and hence benefitted both commercially and socially.

4.5.1 Geographical Information System

'A geographic information system is designed to capture, store, manipulate, analyze, manage, and present all types of spatial or geographical data (refer Fig. 4.6). Geographical Information System is a set of tools for collecting, storing, retrieving, transforming, and displaying spatial data from the real world. In recent times, geographical informational systems are used as a decision support tool especially for planning of investment, planning of infrastructure and economic development' [11].

GIS Components include the following:

- **Hardware**: It includes data storage and digitizers such as scanners and software that turns the colors into lines and filled polygons, large-format printers. It also

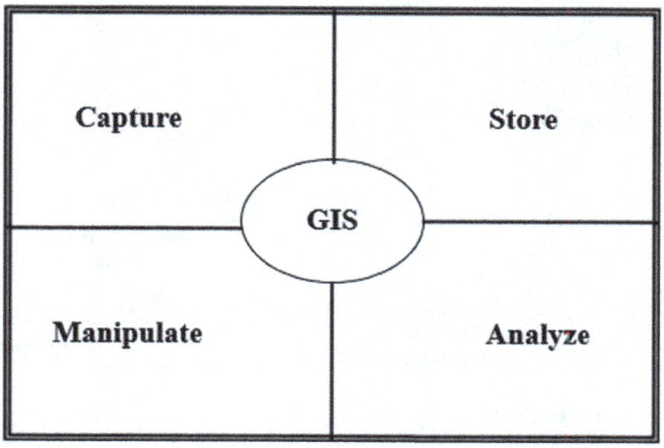

Fig. 4.6 Geographical information system

includes Global Positioning System receivers to collect spatially aware field data for location.

- **Software**: Includes software for Remote Sensing, Map Servers which connect a web user to make use of GIS.
- **Data**: Is a combination of physical features and the associated information which is stored in the tables that are managed by any relational data management system.
- **People**: The end users are the users of GIS system.

Working of GIS is a three-step process:

Visualizing data: The geographic data is stored in the databases that are integrated to the GIS software.
Combine data: It means to put different informative layer on the base map, so that the resultant map is as desired for a particular application. For example, a data scientist needs to analyze vegetation in certain areas of some city, he might take the base map of the city and then add a layer of topographical map on it to identify required areas.
Analyze: It stands for exploring the combined map for the required value or making geographic analysis.

Remote Sensing plays 'an important role in any GIS, and in most cases, allows data to become much more relatable and useful for anyone. Remote sensing technology continues to advance in terms of image resolution and power, thus providing for improved database making GIS more powerful for users to draw insights and intelligence' [12].

Figure 4.7 demonstrates stages in Remote Sensing, the same is explained below:

1. Emission of electromagnetic radiation from Sun.
2. EMR transmission through atmosphere from source to Earth surface (absorption and scattering).

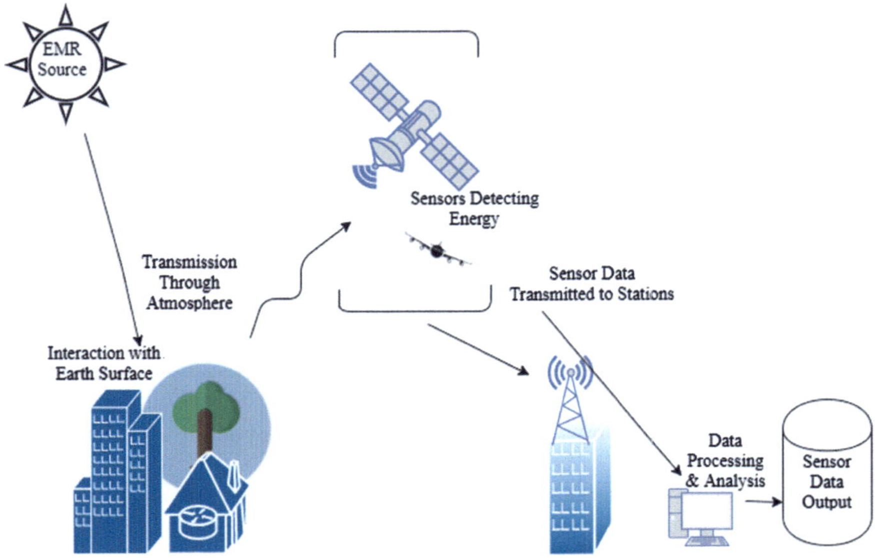

Fig. 4.7 Stages of remote sensing

3. EMR interaction with the suurface of Earth (reflection and emission).
4. Surface to the remote sensor(s) transmission of energy.
5. Sensor data send to ground station for data processing and finally storing output of sensor data.

GIS Techniques: 'GIS tasks range from visualization, monitoring sensors, terrain analysis (modeling of phenomena like flow, radiation, erosion, deposition, infiltration, visibility), to network computations (like shortest paths), surface simplification and working with point cloud data' [13]. Digital Information is used by all GIS technologies, for creating digitized data various data creation methods are utilized while commonly used such method is digitization.

Then this data is represented as graphical information (geographical objects, localization, and topological relationships of these objects) combined with thematic information (descriptors of these objects and their localizations). This is also referred to as 'Duality of Geographical data', as data is stored and represented as geographical features (points/lines/areas/as raster images) and the characteristics of these features. This is done as spatial indices and this spatial data is represented in two ways in GIS:

- Vector representation: 'using points, lines, and polygons and is useful for storing data having discrete boundaries like streets, borders' [13].
- Raster representation: using 'a regular grid of cells/pixels and is useful for storing data that varies continuously like a surface of chemical concentrations,

or an elevation surface' [13]. Aerial photographs are commonly used form of raster images.

Data capture means entering data into the GIS system, consumes most of the time and can be done in a variety of ways. PET film maps (existing paper printed data) can be scanned or digitized. Data can be entered directly using survey tools like GPS (global positioning system). Also, data obtained from remote sensors (satellites, cameras, digital scanners, aerial photographs) can be entered into the GIS system. Once the data is captured, it needs to be edited or corrected for errors or for further processing.

GIS software can also perform data restructuring to convert data to different formats for example converting satellite image to a vector structure.

As earth surface is represented as different models which provide different sets of coordinators viz. longitude, latitude, elevation. GIS can rectify the maps after gathering information from various sources and project them into a single projection using the computational power of computers.

Statistical analysis using GIS: Topological modeling, data modeling, hydrological modeling, network modeling and cartographic modeling can be easily done using GIS software.

GIS Hydrological modeling provides a spatial element by analyzing watershed, slop, aspect etc.. Terrain analysis is elementary to hydrology and involves calculation of slope and aspect. Digital elevation model (DEM, three-dimensional view of landscape) are very useful for hydrological analysis. Network modeling is used for hydrological or infrastructural modeling.

Cartographic modeling is a process in which many thematic layers of same area are generated, managed, and examined.

Map Overlay combines layer of different spatial datasets and to create a fresh vector output.

Geocoding interpolates coordinates with the reference spatial data such as PIN codes. While reverse geocoding is used to obtain the spatial location on the map when ZIP code or PIN code is entered.

Applications of GIS Software: A GIS Software is intended to perform various tasks on almost all types of geographical and spatial data such as storing, retrieving, managing, displaying, and analyzing.

With the help of GIS software maps (and other graphic displays of geographic information) can be produced for study and presentation. GIS software plays a vital role in many application areas like- Environmental Planning and Management, Archeology, Hydrology (Water) Resources, Socioeconomics, Urban Planning, Urban Growth Modeling, Disaster management, Civic planning, Forest Management.

In Fig. 4.8, Analysis of meterological data for 'The Nilgiris' is done using GIS, for the year 2002. The graphical figure shows the temperature (minimum, maximum, avaerage) and rainfall in this geographical region for year 2002.

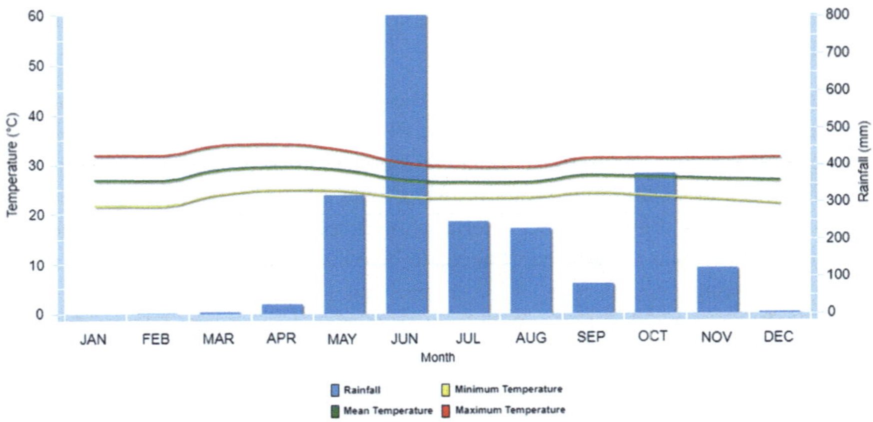

Fig. 4.8 Comprehensive analysis of The Nilgiris weather (rainfall, maximum temperature, minimum temperature, average temperature) for the year 2002

4.5.2 ArcGIS: Experimental Instances

So far in this chapter, we have discussed about the various aspects of Earth Sciences and the enormous possibilities with various Big data analytics platforms available. To further understand about the how we can leverage these available tools both on-premise and on cloud, we will consider ArcGIS and explore few examples on the cloud based version of the tool.

ArcGIS is primarily a 'Geographic Information System' used to work on geological data. ArcGIS is utilized to create maps by compiling geographic information creating maps, analyzing information using maps and storing the information in database. The tool uses information available locally or open database available on web through government portal and community resources.

ArcGIS is a suite of software comprising ArcReader, ArcGIS (earlier ArcView), ArcEditor and ArcInfo and is illustrated in Fig. 4.9.

- ArcReader is used to view and query maps
- ArcGIS for Desktop, which has 3-license level.

  1. ArcGIS for Desktop Basic (earlier ArcView), allows to view spatial data, to do spatial analysis and making layered maps.
  2. ArcGIS for Desktop Standard (earlier ArcEditor), provides advanced tools for manipulation of geodatabases and shapefiles besides the Basic licensing functionality.
  3. ArcGIS for Desktop Advanced (earlier ArcInfo), provides data manipulation functionalities like editing and analysis.

Fig. 4.9 ArcGIS suite—cloud version view

To get started for free, we can create an account on the ArcGIS portal (www.arcgis.com), which provides access to use the platform and assorted case studies loaded with sample maps to make the learning interesting.

Step 1: After registering on the portal and configuring the organization, the portal reads the location information. There is a rich set of repositories available for understanding the platform and maps.

Step 2: Choose a map to work in the dashboard, decide on the parameters required to enrich the map to draw analytics from the map as shown in Fig. 4.10.

Step 3: Identify the geographical area to render the map for drawing insights (Fig. 4.11).

Fig. 4.10 User dashboard for creating map in ArcGIS

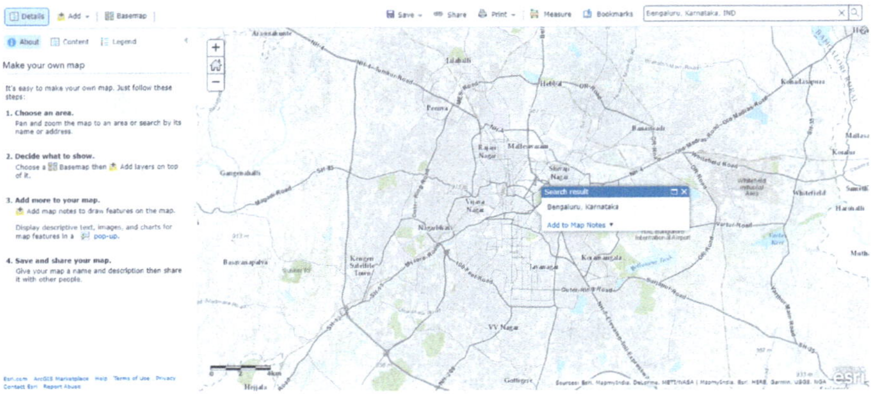

Fig. 4.11 Demonstration of choosing area (location) for map in dashboard

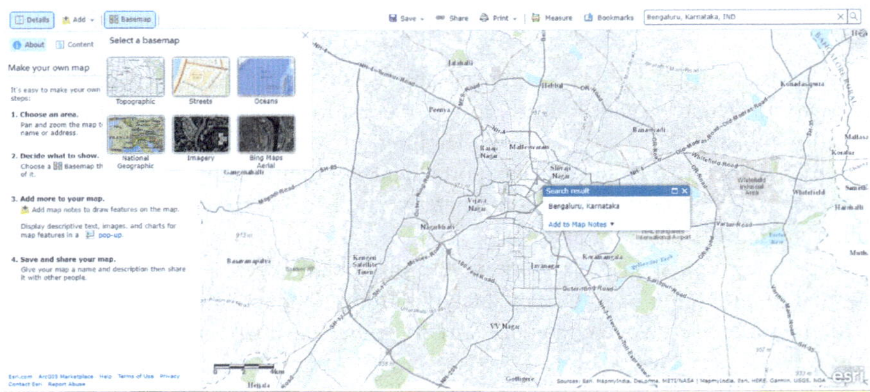

Fig. 4.12 Selecting base map

Step 4: You can also choose from the gallery of contents or choose your own map to start working with the project. This is a base map like topographical, imagery, streets etc. as shown in Fig. 4.12. And later a layer of another map can be added to this selected location for further application based analysis.

In Fig. 4.13 we are using water resources information system to understand Water resources in India. Further we have used the data point available on Surface water bodies and Ground Water resources to do temporal analysis to get surface profile across the map. The snapshots detail few highlights from the analysis done across few areas on the map for the given data points.

The quality of ground water is dependent on multiple parameters and on the various elements present in the water like Arsenic, Fluoride, sulphate, Potassium, TDS, Calcium, Sodium, Carbonate, SAR, Chlorine among other parameters. The data available provide information on ground water quality according to well

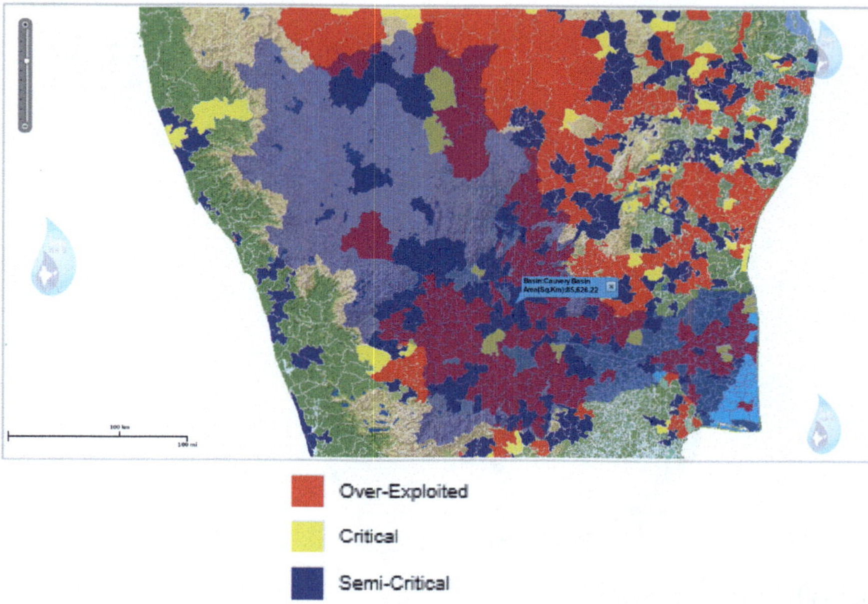

Fig. 4.13 Water resources map created for analysis (highlighting Cauvery basin)

defined area of geographical basins. The information available can also be used to assess the ground water quality changes and compare them in spatial and temporal domain through this info system.

In Fig. 4.14, we have considered the ground water quality for one of the states of India, Karnataka and gradually filtered the choice to a certain area in the urban city of Bangalore.

Figure 4.15, depicts the Temporal analysis of Arsenic affected ground water in the region specified.

The analysis below (Fig. 4.16) provides the Temporal analysis of ground water in Cauvery basin for Fluoride affected ground water.

The data used for analyzing above is drawn from the project "Generation of Database and Implementation of Web Enabled Water Resources Information System in the Country" ('India-WRIS WebGIS').

'India-WRIS is a joint venture of the Central Water Commission (CWC), Ministry of Water Resources, Govt. of India, and Indian Space Research Organization (ISRO), Department of Space, Govt. of India, as per the Memorandum of Understanding (MOU) signed on December 3, 2008 between the two departments for a period of four years—January 2009 to December 2012 and has been extended till December 2015. The data collection, generation and presentation into the portal are continuous activities. The current version India-WRIS WebGIS (Version 4.1) has spatial layers and attributes as per data collected till July

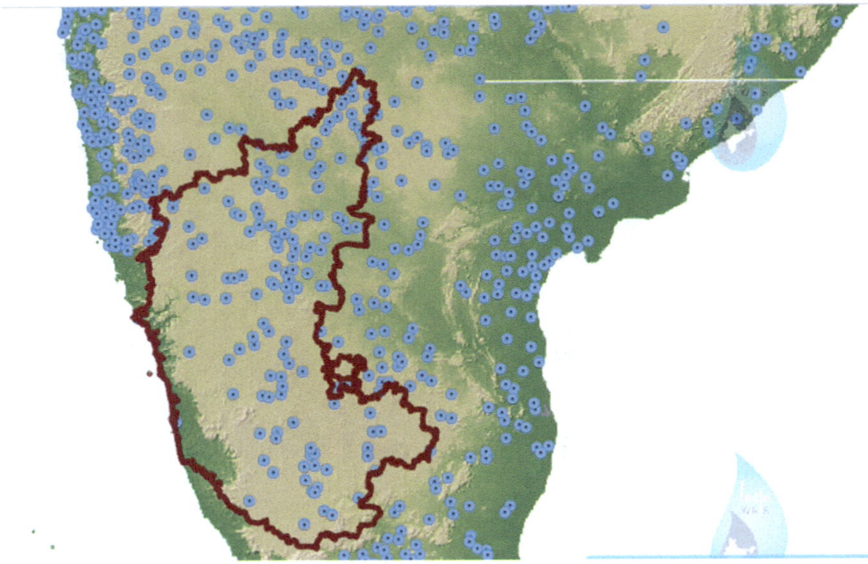

Fig. 4.14 Ground water quality analysis of Karnataka

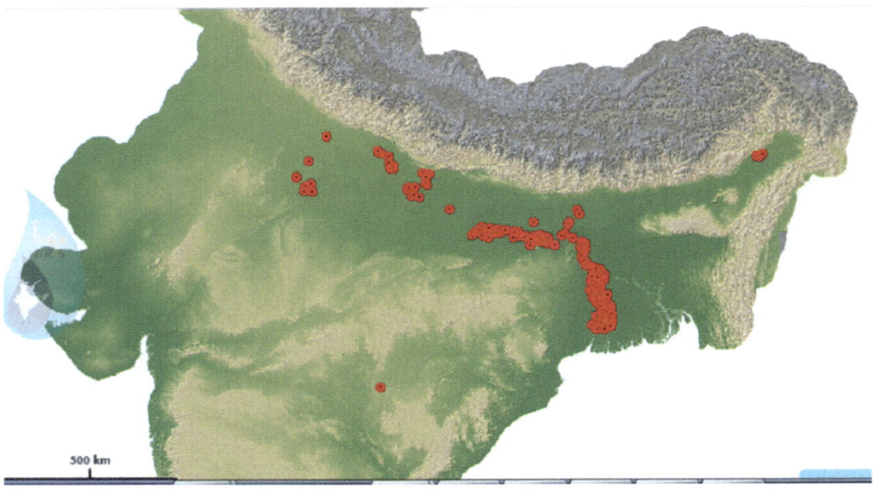

Fig. 4.15 Temporal analysis of arsenic affected ground water

2015. Further updating the attribute data and presentation are being done by the India-WRIS Project Team. These data have been collected from concerned state Govt. departments, CWC offices and Govt. of India departments and are organized in this ArcGIS portal' [14].

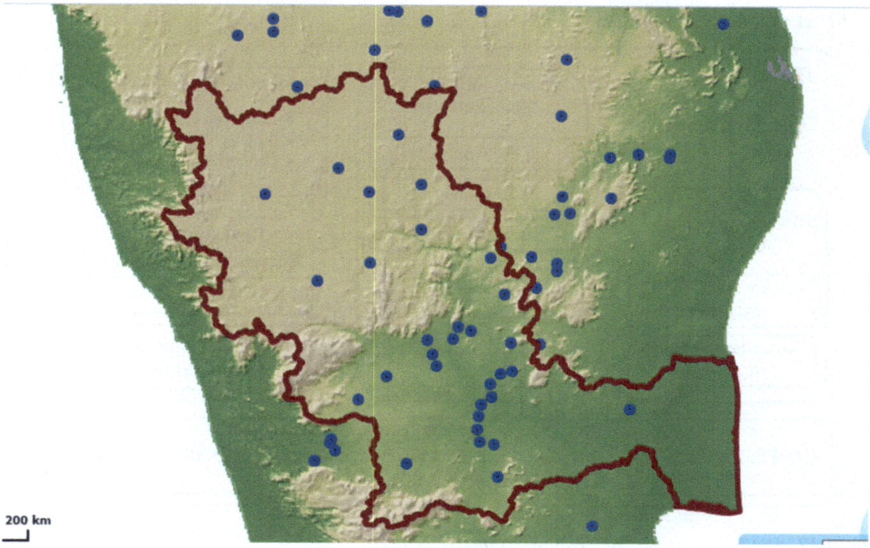

Fig. 4.16 Analysis of ground water in Cauvery basin

4.5.3 IBM InfoSphere BigInsights

IBM BigInsights is offered as a cloud service on IBM Cloud with Hadoop and Spark support. It can be used to analyse structured and unstructured data. 'IBM InfoSphere BigInsights from IBM is a software platform for discovering, analyzing, and visualizing data from disparate sources. We can use this software to help process and analyze the volume, variety, and velocity of data that continually enters the organization every day' [15]. It is a suite of tools which helps users to analyze gigantic volumes of unstructured data and helps in building as well as deploying conventional analytics to capture insights from data. InfoSphere BigInsights provides three ways to work with bigdata and to find the hidden patterns within the data for further use. These scenarios are:

1. Predictive modeling: In predictive modeling patterns from bigdata are analyzed for future predictions helping the clients/users in decision making. For example understanding fraud patterns and fraud reduction in AML system of banking.
2. Consumer sentiment insights: This involves fetching customer/consumer sentiments from social sites, blogs, forums and online surveys for brand management, grieviance redressal and promotions management. In such scenarios, BigInsights text analytics capabilites/tools are used.
3. Research and development: In this scenario, future projects or products are designed and developed. With the availability of BigInsights tool information is collected and analyzed for implementing future projects.

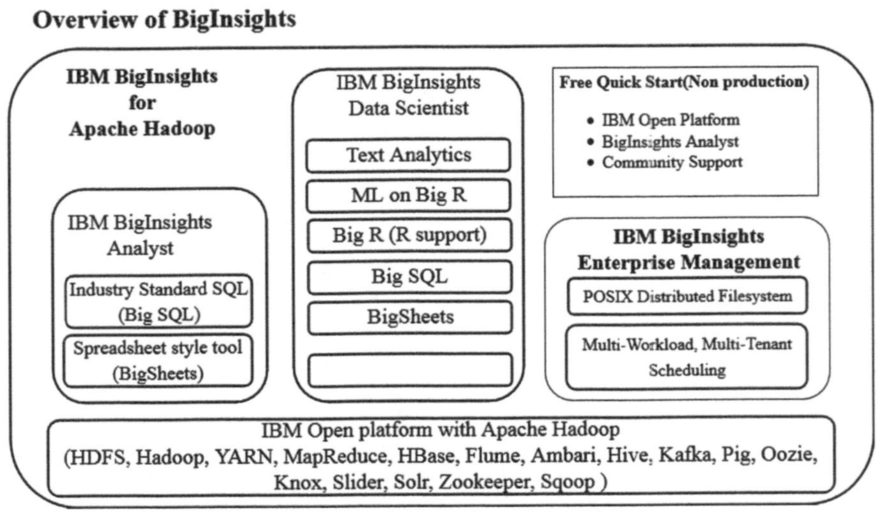

Fig. 4.17 IBM InfoSphere BigInsights suite

As explained in the section above, IBM InfoSphere BigInsights suite (Fig. 4.17) is collection of multiple tools and hence is very powerful solution provided by IBM for big data analytics. BigInsights includes components like BigSheeets, Big SQL, and Text analytics and supports Pig, Hive etc. Table 4.2 describes briefly these components.

4.5.4 *Python*

Python is one among the desired development languages finding favor with developers and enterprises. The strength of Python is its integration of development flexibility, integration of statistical and topical components which makes it a great choice for adoption in data analytics with capability to work with both structured and unstructured data. The capabilities are Python are greatly complemented by largely available third-party packages suitable for analytics for unconventional data.

Table 4.2 Components of BigInsights

BigInsight tool	Brief description
BigSheets	Transform, analyze, model, and visualize big data in a spreadsheet format
Text analytics	Extract structured data from unstructured and semi-structured text using array and powerful visual tool
Big SQL	Turns big data into structured data and run SQL against it. Explores models, analyzes, and visualizes data easily

'Python is a modern, interpreted, object-oriented, open-source language used in all kinds of software engineering. Though it has been around for two decades, it exploded into use in the atmospheric sciences just a few years ago after the development community converged upon the standard scientific packages (e.g., array handling) needed for atmospheric sciences work. Python is now a robust integration platform for all kinds of atmospheric sciences work, from data analysis to distributed computing, and graphical user interfaces to geographical information systems' [16].

The data available in Earth sciences is of varied nature, varying in format, inconsistent across data capture parameters and difference is resolutions of spatial and temporal data. Python in Earth sciences not only provides for exploring Patterns and Predicting and Prescribing actions but is also appropriate to prepare raw data so that it can be processed.

As a part of Earth Science, GIS format provides information about geographic locations, stored in database such as geodatabase, shapefile, coverage, raster image etc. The choice of representation is between using Points, Lines, Polygons, or Regular grid. The typical process included Loading the data, Pre-processing, or cleaning the data, Processing, Analyzing, and displaying the data.

Why Python: With the expansion of scope for using different languages particularly suited for data analysis, the choice of language is still application centric. Each language comes with its own pros and cons, however while deploying certain language for analysis of Earth science data, we need to consider few aspects including processing time, support for disparate tools, third party libraries and appropriateness for use-case.

Consider an example of writing bootloader program, C language would be more apt because it can enable stack, compile, and execute. However, for a GUI driven application, Java can be a good choice because of great built in libraries, being object oriented, wrapper classes, byte code and similar reasons. Python is a great choice for programming language because of its rich third-party libraries. The Python code though runs slower compared to Java or C, but it is a great language to "Glue" various functionalities for stitching a highly complex Earth Science data analytics solution that can draw analytics from Earth sciences data comprising complex data and disparate third-party libraries. Python has powerful polymorphic data types, which do not need declaration unlike other languages and provides for Run-time typing.

However, one of the major differences between these languages is about their runtime and how these languages are compiled. Python is interpreted at runtime and hence isn't compiled until runtime, both C and C++ are compiled to native executables on the other hand Java compiles to JVM bytecode and the Java Virtual Machine compiles the bytecode at runtime.

Based on the way these languages are compiled, they are different in their runtime speed. Of all the languages discussed including C, C++, Java, and Python, both C & C++ execute much faster than Python or Java, because they are compiled to machine readable executables directly. Java is compiled to an intermediary form that is optimized and then runs on a JIT compiler that further optimizes the

Table 4.3 Comparing Python and Java

Java	Python
Is statically typed, therefore all variable names must be explicitly declared with their data types. Attempting to assign an object of the wrong type to a variable name triggers a type exception	Is dynamically typed, assignment statement binds a name to an object, and the object can be of any type, therefore Python does not require explicit data type declaration. Same name can be assigned to objects of different type
In Java, container objects like *ArrayList and Vector*, cannot hold primitives (like *int*) instead can only carry generic type *Object*, however for storing an (*int*) in a Vector, it needs to be converted to an *Integer*. While retrieving an object, it must cast to desired type explicitly	In Python, container objects like lists and dictionaries, can hold objects of any type. While retrieving an object, there is no need to cast it
Java is Verbose language; therefore, it requires more programming constructs to achieve a desired functionality, therefore is not compact	Python is Concise, meaning it requires lesser programming constructs to achieve same functionality and hence is compact language. It also translates in less programming effort while coding in Python and therefore more developer productivity
In terms of Java runtime, the Java code is compiled to an intermediary form that is optimized and then runs on a JIT compiler that further optimizes the bytecode	In terms of Python runtime, the Python code is compiled from source to machine code on the fly as the program executes
Java is more suited for programming GUI Application, business applications, client-side programming, mobile development, server-side programming, web applications	Python is apt scripting language for Web Application, coding artificial intelligence and scientific computing applications, Data Science programming

bytecode. Python is probable the slowest in execution, because it should be compiled from source to machine code on the fly as the program executes. However, the tradeoff between these programming languages is that a programmer can typically write a Python program faster than they can write a Java or other language program. Refer Table 4.3 for comparison of Java and Python.

Python itself is just a construction set and without packages it is only a core, to be productive a data scientist needs more comfortable environment. From a data scientist point of view, Notebooks are the smartest way of playing the data science, a data scientist can choose between ipython or Jupyter notebooks to get started using geoJson or a Json for data representation.

The list below is a compilation of few Python packages which assist in data analytics for Earth Science data.

- Scipy is an interconnected system of OSS (open-source software) for science, mathematics, engineering.

- Pandas which is suitable for any kind of data and offers simple and easy (high performance) tools for data structures and analysis.
- Numpy is a library for scientific computing providing fast and efficient numerical array objects for calculations.
- geopandas provides extended geometric types for spatial operations.
- Matplotlib is a great 2D plotting library.
- For 3D visualization, pyRT is a raytrace and image generator.
- Conda is a package manager.
- Dask is a parallel computing library, with availability of GPU computing this has enhanced the analytics in a big way.
- PyToolz is a set of utility functions for iterators, functions, and dictionaries.
- Numba is a compiler for Python code.
- Xarray is used to work with multi-dimensional data in a pandas-like way.
- pyRT package is used to renders 3D images adding shade and reflections. [17]

Besides these there is handful of packages/tools available for visualization of Earth Science data, few of these are listed below

- PixieDust
- ArcGIS
- bokeh mapbox.

4.5.5 Jupyter Notebook

With the wide spread adoption of Jupyter Notebook data analysis and data visualization has been made very easy for developers and analysts. Jupyter notebook is open source web application providing to create and share code, equations, visualizations, and explanatory text in a document. Jupyter notebooks are based on open standards for interactive computing. Notebooks are used for Data cleansing/cleaning, Data transformation, statistical modelling, machine learning, doing numerical simulation besides many others.

Jupyter notebooks currently support around 40 programming languages such as Scala, Python, R etc. The notebooks provides for easy integration of various tools including big-data software tools, such as ApacheSpark, data analysis libraries from R, Python, Scala, to investigate the data using ggplot2, pandas, dplyr, scikit-learn, etc.

Figure 4.18 explains how Jupyter notebooks can bind the code and visualizations in the same documents and hence are a powerful tool for data analysis in Earth sciences.

Fig. 4.18 Representing code binding in Jupyter notebooks

4.5.6 *MATLAB*

Matlab is one of most popular tool for geophysical computation, data analysis and display. Matlab comprises almost thousand routines for higher math and data manipulation. The libraries available for working on graphics in matlab toolboxes are remarkably good. 'Matlab can be applied to solving problems and developing systems involving mathematical computation, data analytics and statistics, signal and image processing, geographical mapping, and more in the scientific and engineering domains' [18].

Figure 4.19 illustrates World map displaying rivers (blue lines), lakes (cyan) using MATLAB simulation tool. Geo-sciences method includes fundamental statistics for uni-variate, bi-variate, and multivariate datasets; time-series analysis; signal processing; spatial and directional data analysis; and image analysis. To implement data analysis methods and techniques in the geosciences with ease and

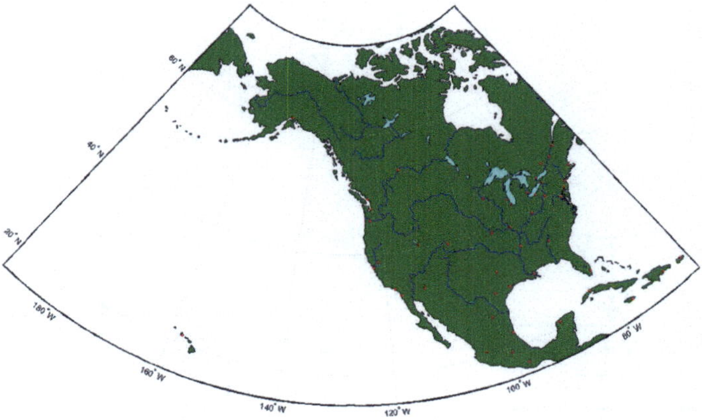

Fig. 4.19 World map displaying rivers (blue lines), lakes (cyan) using MATLAB

pace, Matlab includes very powerful and useful set of toolboxes like statistics toolbox, mapping toolbox, toolbox for wavelets, toolbox for image and signal processing, also it has a wide range of plotting services that can be used with simple commands like *plot ()*.

4.5.7 MapReduce

'MapReduce is a programming model and an associated implementation for processing and generating large data sets. Users specify a map function that processes a key/value pair to generate a set of intermediate key/value pairs, and a reduce function that merges all intermediate values associated with the same intermediate key' [19]. 'It is a potential solution to address the big data challenges as it adopts a more data-centered approach to parallelize runtimes, moving computation to the data instead of the converse. This avoids the movement of large volume data across the network which impacts performance' [20]. 'MapReduce-oriented framework' [21] can be combined with other approaches like K-nearest neighbor based classification to manage big data.

'One of the fundamental processing tasks in the OpenTopography system is generation of DEMs from very dense (multiple measurements per square meter) LIDAR (Light detection and Ranging) topography data. The local gridding algorithm (see Fig. 4.20) utilizes the elevation information from LIDAR measurements contained within a circular search area to compute the elevation of each grid cell. This implementation is data parallel and hence an ideal use case for MapReduce. In the Map phase, input points are assigned to corresponding grid cells (local bins), and in the Reduce phase the corresponding elevations for each grid cell are

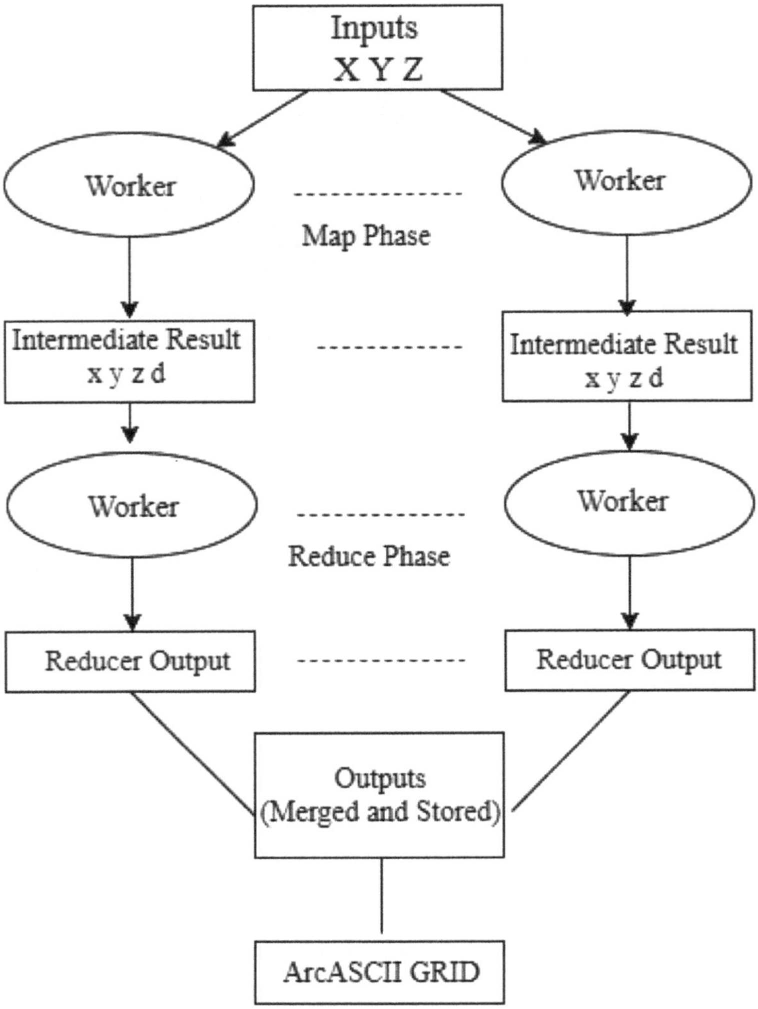

Fig. 4.20 Local gridding algorithm implementation using MapReduce

computed from the local bins. The reduced outputs are merged, sorted and the DEM is generated in the ArcASCII grid format' (also called as ESRI ASCII grid format) [22].

4.6 Conclusions

Earth Science has evolved over the years into a multidisciplinary stream covering aspects from not just typical atmospheric components but also allied sciences like data science, machine learning, deep learning to assist in getting inferences and insights from the vast swathes of data available as part of Earth science.

In this chapter we have explored Earth science with reference to the tools and applications used in exploring Earth science data. With the advances in technology the data available as Earth science is exponentially increasing owing to better resolution, availability of unconventional data sources including unstructured data that have added new dimension to exploring and understanding Earth Science. Further with advent of integrated platforms like ArcGIS, IBM BigInsights etc., data analytics has become easy and more meaningful. We are witnessing explosion of data and hence un-earthen of Earth Science secrets will be instrumental in assisting mankind in preservation and prevention of future calamities by drawing predictions and patterns from the available Earth sciences data.

References

1. Karpatne A, Liess S (2015) A guide to earth science data: summary and research challenges. IEEE Comput Sci Eng 14–18
2. Kempler S, Mathews T (2016) Earth science data analytics tools, techniques and more. In: ESIP Summer Meeting. ESIP Commons
3. Number of earthquakes by year. http://www.johnstonsarchive.net/other/quake1.html
4. Torahi AA, Rai SC (2011) Land cover classification and forest change analysis, using satellite imagery—a case study in Dehdez Area of Zagros Mountain in Iran. J Geogr Inf Syst 3:1–11
5. Steinbach M, Tan P-N, Boriah S, Kumar V, Klooster S, Potter C (2006) The application of clustering to earth science data: progress and challenges. In: Proceedings of the 2nd NASA data mining workshop
6. Clustering. Japan Association of Remote Sensing. Available at http://wtlab.iis.u-tokyo.ac.jp/~wataru/lecture/rsgis/rsnote/cp11/cp11-3.htm
7. Freitas AA (2008) A review of evolutionary algorithms for data mining. In: Maimon O, Rokach L (eds) Soft computing for knowledge discovery and data mining. Springer, New York, pp 79–111. ISSN 978-0-387-69935-6, https://doi.org/10.1007/978-0-387-69935-6_4
8. Engelbrecht AP (2007) Computational intelligence: an introduction, 2nd edn. Wiley, Sussex
9. Freitas AA (2003) A survey of evolutionary algorithms for data mining and knowledge. In: Ghosh A, Tsutsui S (eds) Advances in evolutionary computing: theory and applications. Springer, New York, pp 819–846
10. Cuddy SJ, Glover PWJ (2002) The application of fuzzy logic and genetic algorithms to reservoir characterization and modeling. In: Wong P, Aminzadeh F, Nikravesh M (eds) Soft Computing for reservoir characterization and modeling. Springer, Berlin, pp 219–242. ISSN 14349922
11. http://grindgis.com/what-is-gis/what-is-gis-definition
12. Patra P (2011) Remote sensing and geographical information system (GIS). Assoc Geogr Stud 1–28
13. Algorithms in GIS. Available at http://www.bowdoin.edu/~ltoma/teaching/cs3225-GIS/fall16/index.html

14. India WRIS. Available at http://www.india-wris.nrsc.gov.in/WRIS.html
15. Introduction to InfoSphere BigInsights. IBM Knowledge Center, available at https://www.ibm.com/support/knowledgecenter/SSERCR_1.0.0/com.ibm.swg.im.infosphere.biginsights.product.doc/doc/c0057605.html
16. Lin JW-B (2012) Why Python is the next wave in earth sciences computing. Bull Am Meteor Soc 93(12):1823–1824. https://doi.org/10.1175/BAMS-D-12-00148.1
17. Groenendijk M (2017) Mapping all the things with Python. IBM Watson Data Lab. Available at https://medium.com/ibm-watson-data-lab/mapping-all-the-things-with-python-1228187dc665
18. Kempler L. Teaching with Matlab. Available at https://serc.carleton.edu/NAGTWorkshops/data_models/toolsheets/MATLAB.html
19. Dean J, Ghemawat S (2008) MapReduce: simplified data processing on large clusters. Commun ACM 51:107–113
20. Li Z, Yang C, Jin B, Yu M, Liu K, Sun M, Zhan M (2015) Enabling big geoscience data analytics with a cloud-based, MapReduce-enabled and service-oriented workflow framework. PLoS ONE 10(3):e0116781. https://doi.org/10.1371/journal.pone.0116781
21. Kamal Sarwar, Ripon SH, Dey N, Ashour AS, Santhi V (2016) A MapReduce approach to diminish imbalance parameters for big deoxyribonucleic acid dataset. Comput Methods Programs Biomed 131:191–206
22. MapReduce for Gridding LIDAR Data. In: Applications and Limitations of MapReduce. Available at http://mapreduce-specifics.wikispaces.asu.edu
23. Ericson G, Franks L, Rorer B (2017) How to choose algorithms for Microsoft Azure Machine Learning. Available at https://docs.microsoft.com/en-us/azure/machine-learning/machine-learning-algorithm-choice
24. Ferguson M (2012) Architecting a big data platform for analytics. Intelligent Business Strategies Limited
25. Dutt V, Chaudhry V, Khan I (2012) Pattern recognition: an overview. Am J Intell Syst 2 (1):23–27. https://doi.org/10.5923/j.ajis.20120201.04

Chapter 5
Retrieval of Urban Surface Temperature Using Remote Sensing Satellite Imagery

Jinxin Yang, Man Sing Wong and Hung Chak Ho

Abstract Remote sensing observations provide local, regional, and global information in a holistic view as well as large spatial coverage. With the advancement of remote sensing technology, more instances of where remotely sensed data are used recently to investigate the terrestrial processes and global climate due to their high spatial and temporal resolution. In this regard, there are more studies using remotely sensed imagery for investigation of the Surface Urban Heat Island (SUHI) phenomenon and retrieval of urban surface parameters, e.g. surface temperature, surface albedo, energy fluxes. However, the complex geometric characteristics in urban areas pose great challenges for these retrievals. This chapter presents the Urban Surface Temperature (UST) retrieval with consideration to the urban geometric characteristics in different seasons, analyzing the effective emissivity and urban surface temperature. Emissivity is crucial for surface temperature retrieval. However, the cavity effects and thermal heterogeneity caused by complex buildings affects the effective emissivity over urban areas. In this study, the effective emissivity from ASTER products in different seasons were collected to study the thermal heterogeneity effects on the applications of Temperature and Emissivity Separation (TES) algorithm on the UST retrieval in Hong Kong. Thermal images of Landsat 5 in different seasons were collected for analyses, in which the retrieved USTs, with and without considerations to geometric effects, were compared and analyzed. Finally, SUHI estimates based on two sets of USTs and its impacts on SUHI intensity estimation at different seasons were also studied.

Keywords Seasonal effects · Remote sensing · Urban geometry
Urban surface temperature

J. Yang · M. S. Wong (✉) · H. C. Ho
Department of Land Surveying and Geo-Informatics,
The Hong Kong Polytechnic University, Kowloon, Hong Kong
e-mail: lswong@polyu.edu.hk

J. Yang
e-mail: Jinxin.yang@connect.polyu.hk

N. Dey et al. (eds.), *Big Data for Remote Sensing: Visualization, Analysis and Interpretation*, https://doi.org/10.1007/978-3-319-89923-7_5

5.1 Introduction

Due to rapid urbanization, the Urban Heat Island (UHI) has become one of the major environmental and social problems, with worsening adverse health effects from heat stress, as well as indirectly via air pollution [19]. UHI has long been investigated by air temperature measured at ground level [22]. However, measurements of air temperature only at discrete locations are difficult to provide holistic information of the spatio-temporal temperature patterns. Land Surface Temperature (LST) observed by satellite is an alternative to explore the Surface UHI (SUHI).

Nowadays, there is an increase in the use of remotely sensed data to investigate terrestrial processes and global climate due to their regional and global coverage [7, 23]. Based on remotely sensed data, land cover and land use changes caused by the development of urbanization on the UHI and SUHI have been studied [4, 5, 28].

There have been several studies using remote sensing data to investigate the surface UHI effect and retrieve the urban surface temperature without consideration to geometric effects [1, 2]. The complex geometric characteristics in urban areas pose great challenges for the retrieval of urban surface parameters. Most of the retrieval methods for Urban Surface Temperature (UST) are developed based on an assumption of homogeneous and flat surface across urban areas, in which it may not match with the real-world situation of compact cities especially cities with rapid urban change and development.

Researches on directional anisotropy of observed radiometric temperature over urban areas showed that geometric characteristics should be included in the interpretation of information measured by remote sensors [10, 11, 22]. Soux et al. [20] developed a model to calculate the surface-sensor-sun relations (SUM), that can be used to predict the thermal directional anisotropy for studying the relationships among surface geometry, sensor observations and solar irradiance. Voogt [21] studied thermal anisotropy in urban areas based on the SUM developed by Soux et al. [20]. These two studies have examined different geometric characteristics on temperature measured by very high spatial resolution airborne thermal data, but neither have considered the complex building geometry of a city in relation to the retrieval of UST especially from low spatial resolution satellite data. Thus, it exists a knowledge gap in the ways to integrate urban geometry with satellite remote sensing data.

In order to solve the geometric effects on UST retrieval from the radiance measured by remote sensors, the Improved Urban Emissivity Model by Sky View Factor (IUEM-SVF) was developed by Yang et al. [24, 26] for the estimation of the effective emissivity. The performance of IUEM-SVF was then evaluated by a numerical micro-climate model, Temperatures of Urban Facets in 3D (TUF-3D) [9] and results showed that the IUEM-SVF had promising accuracy for the effective emissivity estimation under thermal homogeneity conditions [26].

In this chapter, two case studies were conducted to further evaluate the performance of IUEM-SVF. The first case evaluates IUEM-SVF with real observation data of the emissivity products of ASTER (Advanced Spaceborne Thermal Emission and Reflection Radiometer) in different seasons. The emissivity products of ASTER, based on TES algorithm, observed at different seasons were collected since the thermal heterogeneity varies with seasons and it affects the observed emissivity. The TES method used in ASTER products was developed by analysing 86 nature surface emissivity spectra data measured in laboratory [6]. Considering the geometric characteristics are very complex, limitations of the application of TES algorithm on the retrieval of urban surface temperature was also analysed.

The second case evaluates the single channel method to retrieve urban surface temperature based on IUEM-SVF, namely the Urban Exitance Model based on Single Channel Method (UEM-SCM) [26]. The UEM-SCM includes the urban geometric characteristics in the single channel method for the retrieval of UST. Thermal images of Landsat 5 from 2010 to 2011 were also collected to analyse the geometric effects on the retrieval of urban surface temperature at different seasons. The retrieved USTs derived from UEM-SCM were compared with the USTs derived from single channel method without including the urban geometric characteristics. SUHI intensities were estimated based on different USTs and the impacts of different retrieval methods on the SUHI intensity estimation were also evaluated. This case study evaluated the seasonal effects on the application of UEM-SCM in UST retrieval.

The urban areas of Kowloon peninsula and the northern side of Hong Kong Island were selected as the research areas because of its complex urban morphology. Located in the coastal region of southeastern China (22° 17′N, 114° 09′E), Hong Kong is one of densest cities in the world. Areas of interest in this study are the major urban areas of Hong Kong with relatively low elevation of 8 m [3], and were formed with buildings in a compact setting. As the high-rise buildings and narrow streets may prohibit the application of satellite data, it is necessary to explore the geometric effects on the UST retrieval. More information of research areas can be referred to Yang et al. [25].

5.2 Data Used

5.2.1 Satellite Data

Launched on 1st March 1984, Landsat 5 completed its mission in November 2011 (http://landsat.usgs.gov/about_landsat5.php). The original spatial resolution of the thermal band is 120 m, while the distributed Landsat 5 thermal images are re-sampled at 30 m spatial resolution. The revisit time of Landsat 5 is 16 days (http://landsat.usgs.gov/about_landsat5.php). The cloudless thermal images of Landsat 5 from 2010 to 2011 were collected to analyze the urban geometric effects on UST retrieval. Such time period was chosen since the airborne LiDAR data on urban geometry were obtained during the period of December 2010 and January 2011.

USTs were derived from the thermal images of Landsat 5 by applying the single channel method without considering geometric effects and UEM-SCM were used to analyze urban geometric effects on UST retrieval.

The ASTER AST_05 emissivity products acquired on 13th March 2013 and 4th August 2013 were used to evaluate the IUEM-SVF. The data on 13th March 2013 can represent the spring season and the data on August 4th 2013 can represent the summer season. Table 5.1 shows the satellite data used in this chapter.

5.2.2 Urban Morphological and Ancillary Data

Digital Surface Model (DSM) at 1 m spatial resolution and building GIS data were acquired to calculate the Sky View Factor (SVF) [12]. The DSM data were obtained during December 2010 and January 2011. Since the land cover changes normally take time in the urban areas, the DSM data were used to analyze geometric effects on the retrieval of UST from the satellite data shown in Table 5.1. The land use classification data were obtained from the Hong Kong Planning Department. The classification data and building GIS data were applied to estimate the material emissivity of surfaces and building facets. More information about DSM and classification data and building GIS data can be referred to Yang et al. [25].

However, the classes from land cover data and building data may not match with the materials listed in the spectral emissivity library. Therefore, this study is based on the assumption that the classes from the land cover data include multiple materials in the library by averaging the emissivity spectra with assumption that they are associated with each land cover class. The library of Kotthaus et al. [8] provides emissivity spectra of different construction materials over urban areas. Considering that it is difficult to obtain the exact material information of each

Table 5.1 Satellite data list

Satellite	Date	Local time	Bands/product
Landsat 5	2010/01/14	10:37 am	Thermal band (band 6)
	2010/03/26	10:43 am	
	2010/09/18	10:42 am	
	2010/10/29	10:36 am	
	2010/11/30	10:36 am	
	2010/12/23	10:42 am	
	2011/01/01	10:36 am	
	2011/02/02	10:36 am	
	2011/03/13	10:42 am	
	2011/06/01	10:41 am	
ASTER	2013/03/13	10:36 pm	Emissivity products
	2013/08/04	10:36 pm	

Table 5.2 Material emissivities of impervious surfaces at different wavelengths

	Bands	Wall	Roof	Road/street
Landsat 5	Band 6	0.937	0.956	0.956
ASTER	Band 10	0.915	0.907	0.896
	Band 11	0.880	0.906	0.906
	Band 12	0.835	0.899	0.905
	Band 13	0.886	0.945	0.948
	Band 14	0.916	0.947	0.949

building facet, the average material emissivity for each facet was estimated by the material emissivity of general construction materials e.g. concrete, cement or asphalt in the spectral library [24, 25].

The thermal data of Landsat 5 were applied for the retrieval of USTs. The material emissivity of wall facets is 0.937, which was calculated as the average value of emissivities of brick, concrete, and glass in the Band 6 of Landsat 5. The emissivity of the rooftops is 0.956, which was assumed as the mean value of emissivities of roofing shingle, tile, concrete, and cement. The roads were assumed to be constructed by asphalt, brick, concrete, and cement, with the emissivity estimated to be 0.956. Table 5.2 shows the summary of the spectral emissivities used in this study.

The ground station data, including air temperature, relative humidity and ground atmospheric pressure, were collected from Hong Kong Observatory. These data were used as input parameters for atmospheric correction of the thermal bands of Landsat 5. The sea surface temperatures at (22.30472° N, 113.87917° E) and (22.18222° N, 114.30333° E) were collected to validate the atmospheric correction of the thermal bands of Landsat 5.

5.3 Case Studies

5.3.1 Evaluation of IUEM-SVF

Multiple reflection and re-absorption cause of cavity effect over urban areas. It is expected that the emissivity and absorption of the urban pixel in complex environments are higher than the emissivity of the flat surface covered by same material. Based on the thermal radiative transfer, the effective emissivity model of IUEM-SVF [26] can be written as:

$$\varepsilon_i'' = \varepsilon_i' + (1 - \varepsilon_i')(1 - V_i)\varepsilon_i \tag{5.1}$$

where $\varepsilon_i' = \varepsilon_i/(1 - (1 - \varepsilon_i)(1 - V_i))$ [24], ε_i is the material emissivity of pixel i, V_i is the SVF of pixel i.

The performance of IUEM-SVF was evaluated by a micro-climate numerical model (TUF-3D) under the condition of buildings with simple geometric characteristics. The TUF-3D can compute a full radiative and convective for each of the large number of facets used to represent urban objects. This model, therefore, calculates both the object surface temperature and the associated emittance in a generic and accurate way, which are then combined to determine the surface emissivity and to evaluate the IUEM-SVF effective emissivity.

In this chapter, the emissivity products of ASTER AST_05, derived from the TES method [6], were collected to evaluate the IUEM-SVF. The wavelength of five ASTER thermal bands ranges from 8.125 to 11.65 μm. The spectral emissivity of ASTER products is the emissivity over the urban canopy which captures geometric effects in urban areas, thus it is different from the material emissivity. The products of ASTER AST_05 were validated by the measurement of emissivity over water surface with accuracy of ± 0.015 [6]. The TES algorithm uses spectral contrast among the multiple spectra and the empirical relationship between spectral contrast and the minimum spectral emissivity to calculate the spectral emissivity. It then uses the spectral ratio to calculate the emissivity of other bands. Over urban areas, the spectral information obtained by remote sensors is the spectral radiance affected by the geometry and thermal heterogeneity. This affects the spectral emissivity and the contrast thus, TES algorithm captures the geometric effects over urban areas. The emissivities in bands 10–14 from ASTER AST_05 product was used to analyze the relationship between the SVF over urban areas. While the radiative transfer in urban areas is not explicitly taken into account in the TES method, the performance of TES over urban areas were explored by comparison with the IUEM-SVF effective emissivity.

5.3.2 Application with UEM-SCM

For the second case, thermal band 6 of Landsat 5 images were used to retrieve the USTs for analyzing the variation of geometric effects in different seasons. The impact of urban geometry was investigated by comparing USTs derived from the single channel method without including geometric effects with USTs derived by UEM-SVF. The radiance observed by the remote sensors at the Top of Atmosphere (TOA) can be written as [13]:

$$I_i = E_i \tau_i + R_{i,at\uparrow} \tag{5.2}$$

where E_i is the channel radiance of pixel i leaving the urban canopy (W m^{-2} sr^{-1} μm^{-1}), τ_i is the atmospheric transmittance and $R_{i,at\uparrow}$ is the upward thermal radiance of atmosphere (W m^{-2} sr^{-1} μm^{-1}) [13].

When geometric effects are not included, the thermal infrared radiance leaving ground (exitance) can be written as [25]:

$$E_i = \varepsilon_i B T s_i + (1 - \varepsilon_i) * R^{\downarrow}_{i,at} \quad (5.3)$$

where ε_i is the material emissivity of pixel i, B is the Planck function, Ts_i is the surface temperature of pixel i, $R^{\downarrow}_{i,at}$ is the atmospheric downward radiance.

When the effects of geometric characteristics and local topography are included, the thermal radiance leaving urban canopy (E_i) comprises the emittance of facets in the observed built-up space, reflected radiance by the facets within pixel i, and by the neighboring scene elements. Considering the geometric effects, the UEM-SCM developed by Yang et al. [26] is written as:

$$E_i = \varepsilon'_i B(Ts_i) + \left(1 - \varepsilon'_i\right)\left\{(1 - V_i)\varepsilon(i)B(Ts_i) + V_i * R^{\downarrow}_{i,at} + V_{adj,i} * R_{adj}\right\} \quad (5.4)$$

B(Ts_i) is the radiance emitted by the pixel i, $V_{adj,i}$ is the view factor between pixel i and adjacent pixels, R_{adj} is the radiance from adjacent pixels to pixel i. The method of SVF calculation for Landsat data can be referred to Yang et al. [25] and the method of SVF calculation for ASTER data can be referrer to Yang et al. [26].

5.4 Results

5.4.1 *Case Study I" Comparison Between IUEM-SVF and Other Methods of Emissivity Retrieval*

ASTER images acquired at 22:36 local time on 13th March 2013 and 4th August 2013 (Fig. 5.1) were used. The urban areas in Hong Kong's Kowloon peninsula and the northern side of Hong Kong Island were selected. The emissivity over impervious flat surfaces is lower than that over built-up areas as shown in Fig. 5.1. The difference between the emissivity on 13th March 2013 and 4th August 2013 may be due to the thermal heterogeneity which affects radiative transfer within mixed pixels. Thus, the spectral contrast in emissivity would also be varied. This affects the results of the TES algorithm. The emissivity product of ASTER also implicitly includes information of thermal heterogeneity conditions, and the retrieved emissivities of the same location may be different at different times and dates.

Figure 5.2 shows the relationship between the SVF and spectral emissivity of ASTER AST_05 products in the ASTER bands 10–14 (8.125–11.65 μm) over the Kowloon peninsula. Since the material and thermal condition of road surfaces are relatively homogenous, the areas dominated by roads were selected. The effective emissivity in Fig. 5.2 is the mean emissivity of ASTER AST_05 over all pixels with an incremental SVF range of 0.1. The results indicate that the effective emissivities in ASTER thermal bands have negative correlations (Fig. 5.2) with

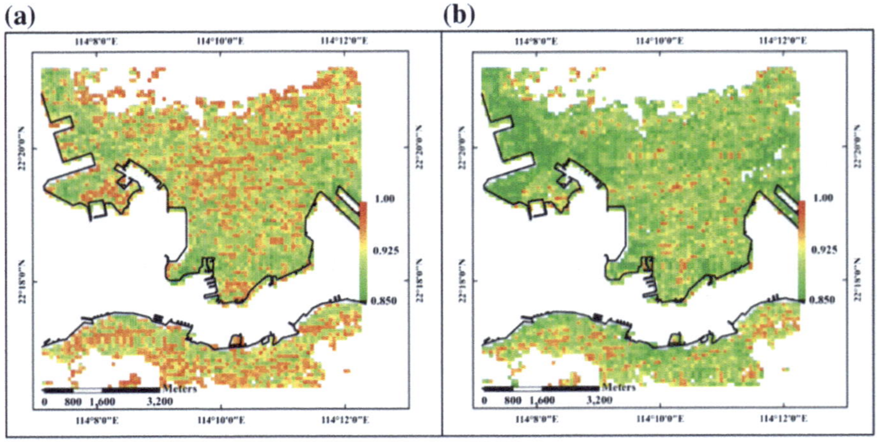

Fig. 5.1 Emissivity products from ASTER AST_05: **a** emissivity of band 12 of ASTER on 13th March, 2013; **b** emissivity of band 12 of ASTER on 4th August, 2013

SVF (r^2 = 0.89, 0.93, 0.89, 0.87 and 0.89 on 13th March 2013, r^2 = 0.86, 0.99, 0.98, 0.93, and 0.86 on 4th August 2013). When the SVF increases, the effective emissivity decreases since the areas with higher SVF capture less radiance and result in lower effective emissivity. When the surface is flat and without surrounding buildings (SVF ~ 1), the radiance emitted by surfaces is directly towards the sky. Thus, the emissivity is close to the material emissivity which is lower than the effective emissivity caused by cavity effects. This results in decreases of effective emissivity with the increases of SVF, and a negative correlation between effective emissivity and SVF (Fig. 5.2). However, the differences between the minimum and maximum values of emissivities of band 13 and 14 on 4th August 2013 are larger than the differences on 13th March 2013. This may be caused by the larger thermal heterogeneity on 4th August 2013 than 13th March 2013. The spectral contrasts of emissivites on 4th August 2013 is approximately 0.02 higher than that on 13th March 2013, which may affect the application of TES in urban areas. The difference between the emissivity on 13th March 2013 and on 4th August 2013 can be explained by the interaction of components within the pixels. Additionally, the geometric effects also change the spectral contrast. This may cause a difference in the TES emissivity, which is based on spectral contrast and ratios, of the same location when the component temperatures are different since the interactions among different components may induce significant changes in the spectral contrasts and ratios.

In order to explore the geometric effects on the application of TES algorithm over urban areas, the IUEM-SVF was used to calculate the effective emissivities which are affected by cavity effects. Figure 5.3 shows the emissivity of ASTER Band 13 from ASTER emissivity products and the effective emissivity estimated by the IUEM-SVF. The emissivity based on IUEM-SVF is higher than the emissivity of ASTER products, while the correlations between the modelled and ASTER

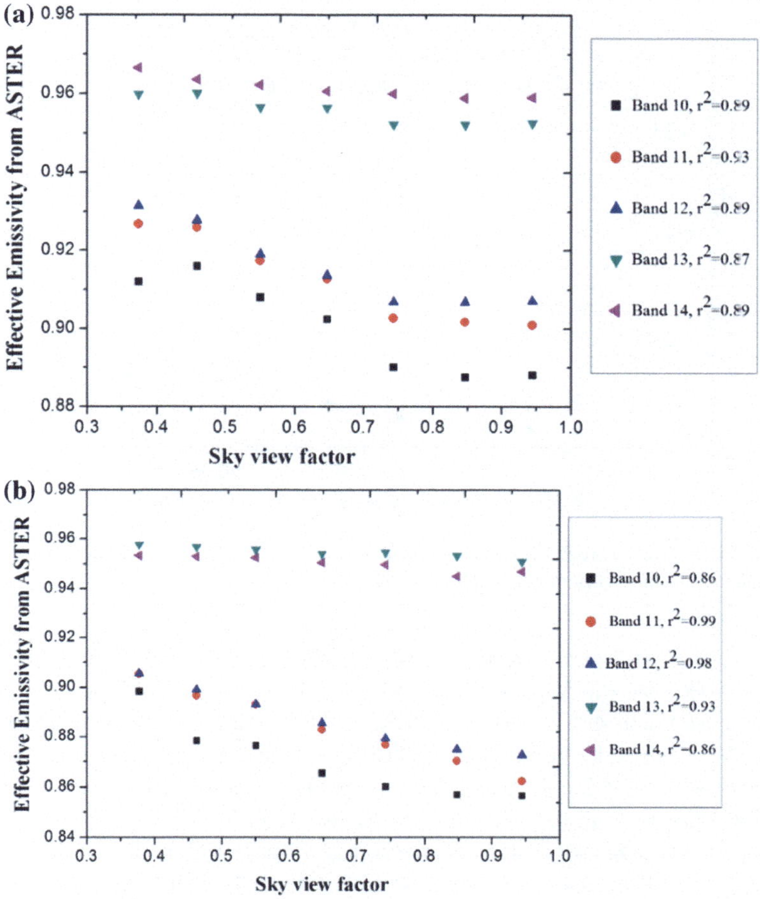

Fig. 5.2 Relationship between the effective emissivity given by the ASTER products and the sky view factor: **a** 13th, March, 2013; **b** 4th, August, 2013

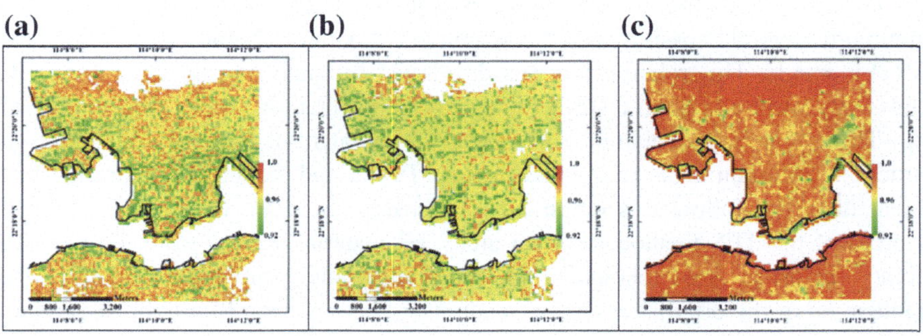

Fig. 5.3 Emissivity of ASTER band 13 derived from ASTER AST_05 products and IUEM-SVF: **a** 13th, March, 2013; **b** 4th August, 2013; **c** IUEM-SVF

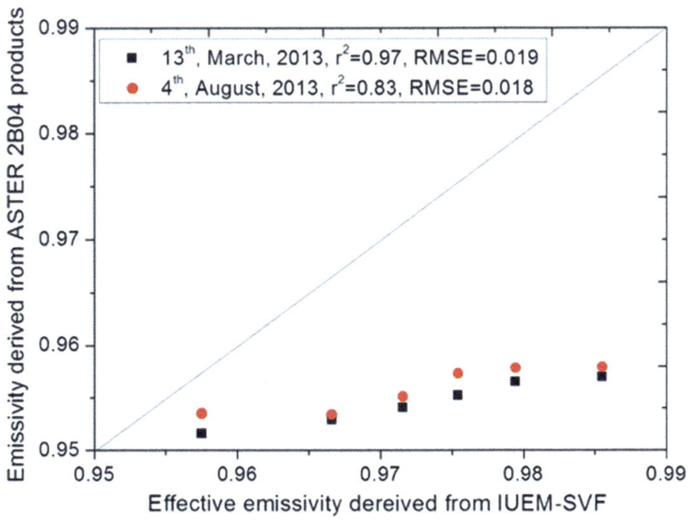

Fig. 5.4 Relationship between effective emissivity from the IUEM-SVF and ASTER emissivity data products in the road-dominated areas

emissivity were still high over the road-dominated areas (r^2 = 0.97 on 13th March 2013 and 0.83 on 4th August 2013) and the RMSEs were 0.019 on 13th March 2013 and 0.018 on 4th August 2013 (Fig. 5.4). The emissivity estimated by the IUEM-SVF tends to be higher than ASTER emissivity, and the deviation increases with decreasing SVF, i.e. the dependence on geometry described by the IUEM-SVF model is different from the one captured by the ASTER TES algorithm.

One of the reasons why effective emissivities from IUEM-SVF were higher than that from ASTER products is that the ASTER emissivity is affected by the thermal and material heterogeneity. Another reason is that the material emissivity used in IUEM-SVF is not perfectly accurate since it is based on the approximation of classification and building data, and it is impossible to obtain accurate material information for the whole urban areas.

One error source for the TES algorithm is the empirical relationship between the minimum spectral emissivity and the Minimum-Maximum Deviation (MMD) in the ASTER bands. This relationship was developed by analyzing 86 laboratory TIR emissivity spectra of natural surfaces, e.g. rocks, vegetation, snow, soils, and water [6]. These materials of natural surfaces are different from materials of built constructions in urban areas (e.g. glass, metal). Payan and Royer [17] indicated that metal does not follow this empirical relationship. In order to explore the error sources of the TES method in urban areas, 67 material spectra from the material library for urban areas (excluding the metal materials) were used to analyze the application of the original empirical relationship between the Minimum-Maximum Deviation (MMD) and the minimum emissivity ($\varepsilon_{min} = 0.994 - 0.687 * MMD^{0.737}$) for urban building construction materials. The results in Fig. 5.5 demonstrate that

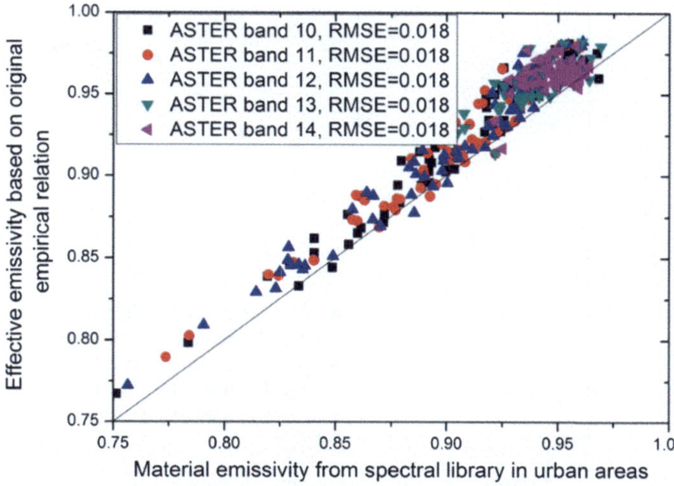

Fig. 5.5 The relation between the urban emissivity based on the original empirical relationship $\varepsilon_{min} = 0.994 - 0.687 * MMD^{0.737}$ versus the actual values in the ASTER spectral library

the original empirical relationship can still estimate the material emissivity over urban areas with RMSE value of 0.018. When eight metal materials were adopted to estimate the emissivity based on $\varepsilon_{min} = 0.994 - 0.687 * MMD^{0.737}$, the RMSE increased to 0.297 (Fig. 5.6). Since increasing amounts of metals are used in urban building constructions, this can induce significant errors.

Additionally, the geometric effects and the thermal and material heterogeneity can induce significant variation in spectral contrast. The complex effects of thermal and material heterogeneity were discussed in Yang et al. [26]. Geometric effects cause the effective spectral emissivities in high density areas close to 1 and this changes the spectral contrast. Thus, the original relationship between spectral contrast and the minimum emissivity is neither accurate nor reliable over urban areas, while the ASTER TES algorithm still adopts the original empirical relationship to estimate the urban emissivities. In order to explore the implications of geometric effects on the application of TES algorithm over urban areas, the effective spectral emissivities were calculated from IUEM-SVF combining with material emissivity spectra under different geometric conditions (SVF = 0.3, 0.4…0.8). The original empirical relationship between the spectral contrast and minimum emissivity was used to calculate minimum emissivity and the corresponding spectral emissivities in the ASTER bands. Figure 5.7 shows the comparison between the effective emissivity based on IUEM-SVF and the ASTER emissivity based on the original empirical relation between the minimum emissivity and the MMD ($\varepsilon_{min} = 0.994 - 0.687 * MMD^{0.737}$). The results show that the emissivities based on the original empirical relationship are smaller than that based on the IUEM-SVF. The difference between these two sets of emissivities becomes smaller with increasing SVF (over flatter and more open surfaces).

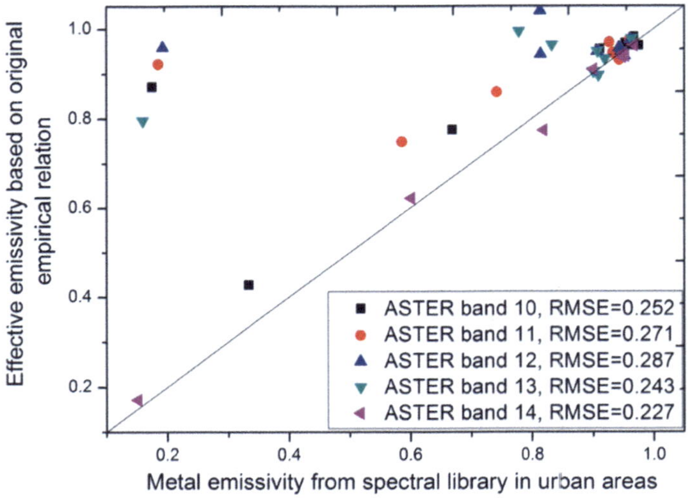

Fig. 5.6 The relation between the metal emissivity and based on the original empirical relationship $\varepsilon_{min} = 0.994 - 0.687 * MMD^{0.737}$ (the x-axis is the metal emissivity based on the ASTER spectral response function and metal spectral function, y-axis is the re-calculated metal emissivity based on the original empirical relationship $\varepsilon_{min} = 0.994 - 0.687 * MMD^{0.737}$ and the spectral emissivity of metal)

According to Oltra et al. [16], the retrieved emissivities based on the TES method were higher than the material emissivities because of the multiple reflections over urban areas. This indicates that the TES algorithm can be applied to calculate the effective emissivity affected by complex building geometries over urban areas because remote sensors can capture the actual spectral radiance. The problem is that the contrast among the actual spectral radiance observed by remote sensors over urban areas does not follow the original empirical relationship of ε_{min}-*MMD*. Urban geometric effects induce the changes of spectral shape and contrast of the spectral radiance. Thus, it is recommended to build a new empirical relationship including the parameters of different urban geometric characteristics and surface materials.

5.4.2 Case Study II: Surface Temperature Retrieval with UEM-SCM

In order to explore the variation of geometric effects on the UST retrieval in different seasons, thermal images of Landsat 5 from 2010 to 2011 were used to retrieve the USTs based on single channel methods with and without consideration to geometric effects.

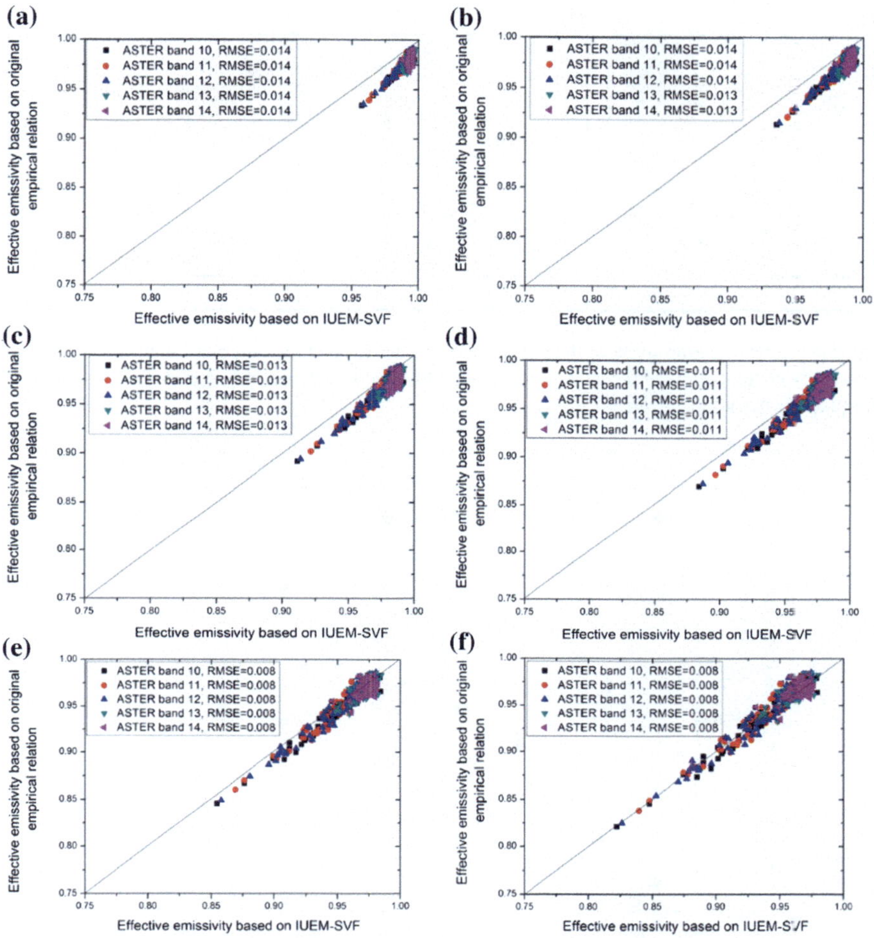

Fig. 5.7 Comparison of the effective emissivity based on the IUEM-SVF with the emissivity based on the original empirical relationship $\varepsilon_{min} = 0.994 - 0.687 * MMD^{0.737}$ under different geometric conditions: **a** svf = 0.3; **b** svf = 0.4; **c** svf = 0.5; **d** svf = 0.6; **e** svf = 0.7; **d** svf = 0.8

Figure 5.8 shows different estimations of emissivity in the band 6 of Landsat 5. The effective emissivity is much higher than the material emissivity of the horizontal surface and complete surface, especially over the built-up areas. The average difference between the effective emissivity derived from IUEM-SVF and the material emissivity without considering geometric effects is 0.016, while the difference over built-up areas is about 0.02–0.03.

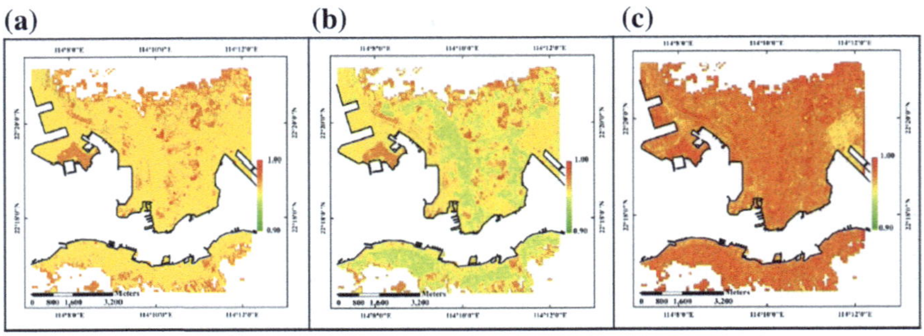

Fig. 5.8 aMaterial emissivity of the flat surface within the pixel in band 6 of Landsat 5; **b** material emissivity of the complete facets within the pixel of band 6 of Landsat 5; **c** effective emissivity of the pixel in the Landsat 5 band 6

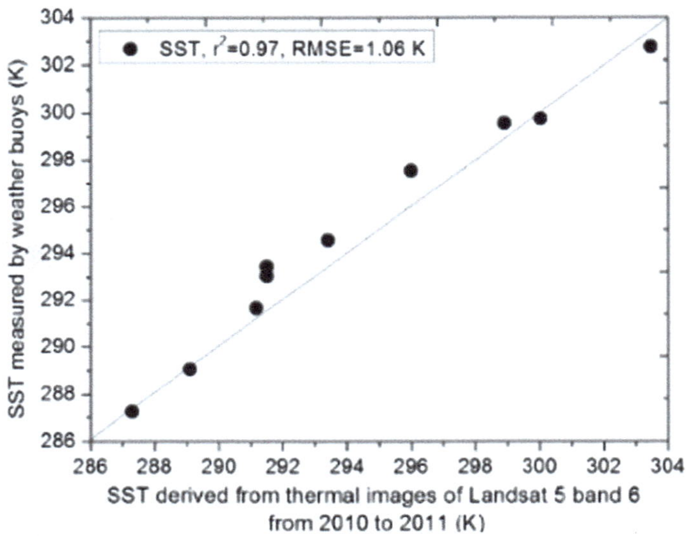

Fig. 5.9 Comparison between the SST derived from thermal images and the SST measured by weather buoys: SST derived from single channel method from thermal images of Landsat 5

5.4.2.1 Atmospheric Correction for UST Retrieval

The data of Sea Surface Temperature (SST) from 2010 to 2011 and 2013 at the time of satellite passing over Hong Kong were collected to validate the atmospheric correction of the thermal images. Figure 5.9 shows the comparison between the SST derived from thermal images and the SST measured by weather buoys. The results indicated that the atmospheric correction was accurate and stable for the single channel method for Landsat 5 from 2010 to 2011 ($r^2 = 0.97$ and RMSE = 1.06 K).

5.4.2.2 Evaluation of UEM-SCM

Figure 5.10 illustrates the USTs retrieved based on the single channel method of Landsat 5 between 2010 and 2011. The left column of Fig. 5.10 is the USTs without considering geometric effects and right column is the USTs derived from UEM-SCM. The USTs derived from UEM-SCM are lower than the USTs without considering geometric effects. This is because the material emissivity does not take multiple reflection and cavity effect into account and neglects the interaction among component facets within the pixel. This underestimates the emissivity effect over urban areas and overestimates the USTs.

As identified by Figs. 5.11 and 5.12, the differences between the UST without considering geometric effects and UST derived from UEM-SCM over built-up areas are the largest (i.e. can reach 2.5 K in winter and late autumn) and the differences in vegetation-dominated areas are the lowest (i.e. less than 1 K). The areas without buildings are also affected by the neighboring buildings, thus the USTs derived from UEM-SCM are also lower than the USTs without considering geometric effects over the road-dominated areas, with the mean differences ranging from 1–1.2 K. The difference between the UST without considering geometric effects and UST derived from UEM-SCM is less than 0.5 K over the flat areas where it is far away from surrounding buildings, e.g. Kai Tak Cruise Terminal, which is the least affected by geometry.

The differences between the UST without geometric effects and USTs derived from UEM-SCM also presents a seasonal pattern (Fig. 5.11). Figure 5.11 indicates that the differences in winter (i.e. December and January) are higher than in summer and autumn (i.e., June 1st, 2011, September 18th, 2010), and this is mainly due to the different atmospheric downwelling radiances. The atmospheric downwelling radiance in summer is higher than in winter over mid-latitude tropical areas.

When geometric effects are taken into account, the urban areas can absorb more and reflect less atmospheric downwelling radiance hence USTs derived from UEM-SCM increases. When urban areas are assumed as a flat surface, the urban areas reflect more atmospheric downwelling radiance. The USTs without considering geometric effects are higher than those derived from UEM-SCM because of the underestimation of the emissivity. However, the high atmospheric downwelling radiance in summer makes a smaller difference between the USTs derived from UEM-SCEM and USTs without considering geometric effects.

Figure 5.13 presents the relationship between average differences of two sets of USTs and the atmospheric downwelling radiance. When the atmospheric downwelling radiance is higher, the average difference between USTs derived from UEM-SCM and USTs without considering geometric effects is smaller. Thus, the difference between the USTs derived from UEM-SCM and the USTs without considering geometric effects over urban areas is even smaller in summer when the thermal heterogeneity is not taken into account. The retrieved USTs derived from UEM-SCM presents more detailed spatial patterns over built-up areas. If geometric effects are not included, the surface temperatures of streets or roads can be

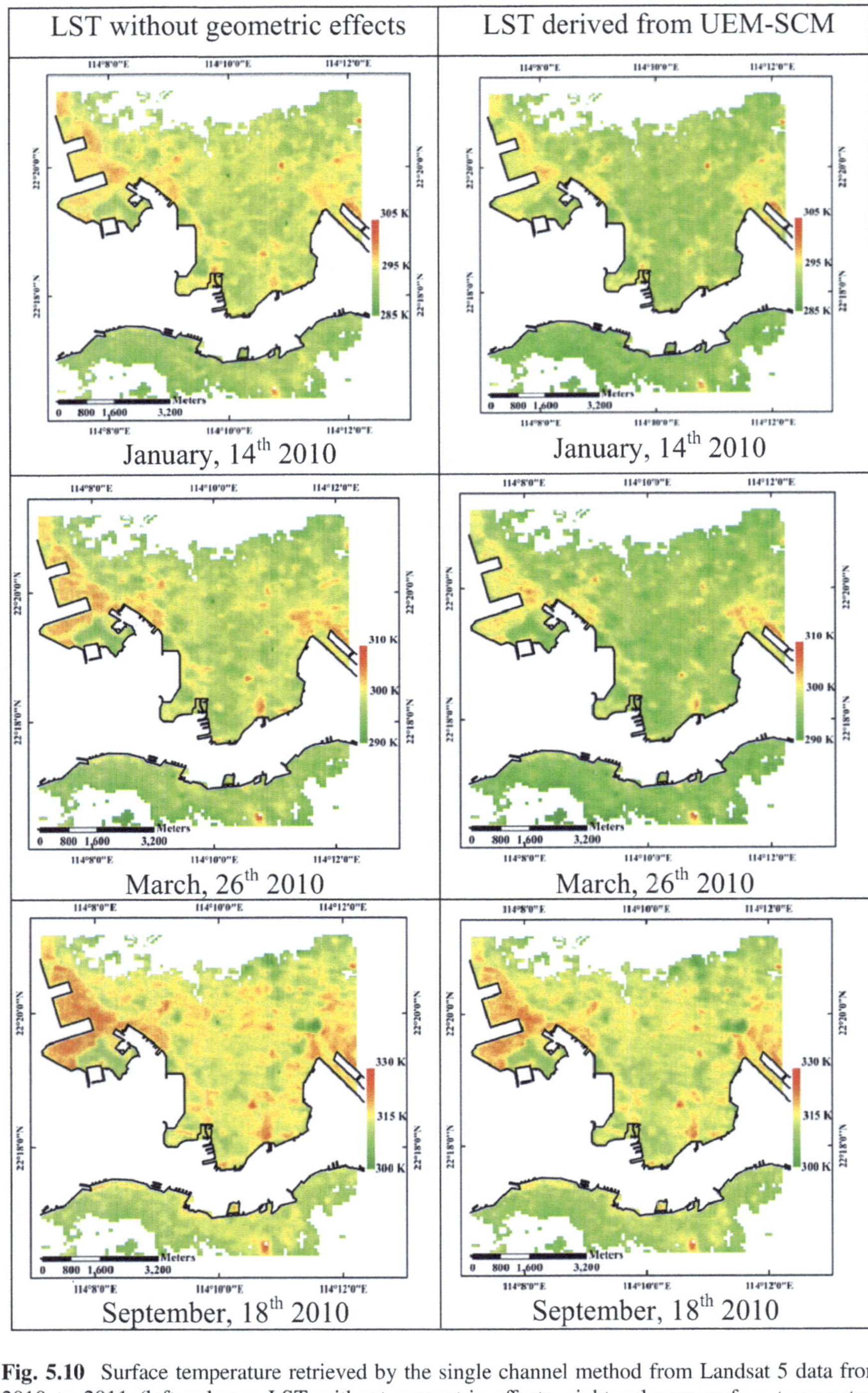

Fig. 5.10 Surface temperature retrieved by the single channel method from Landsat 5 data from 2010 to 2011 (left column: LST without geometric effects; right column: surface temperature derived from the UEM-SCM). (Note that the ranges of the legends are different)

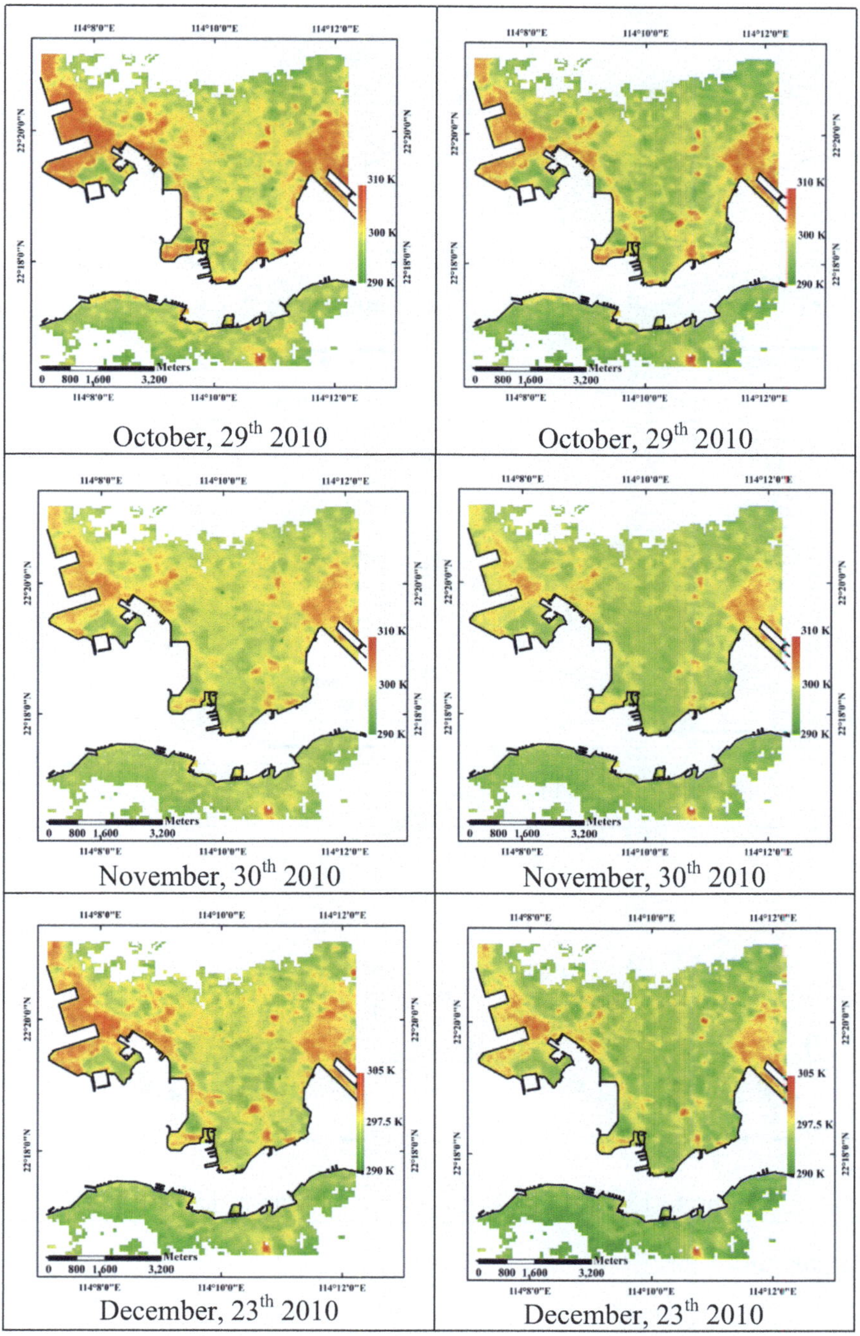

Fig. 5.10 (continued)

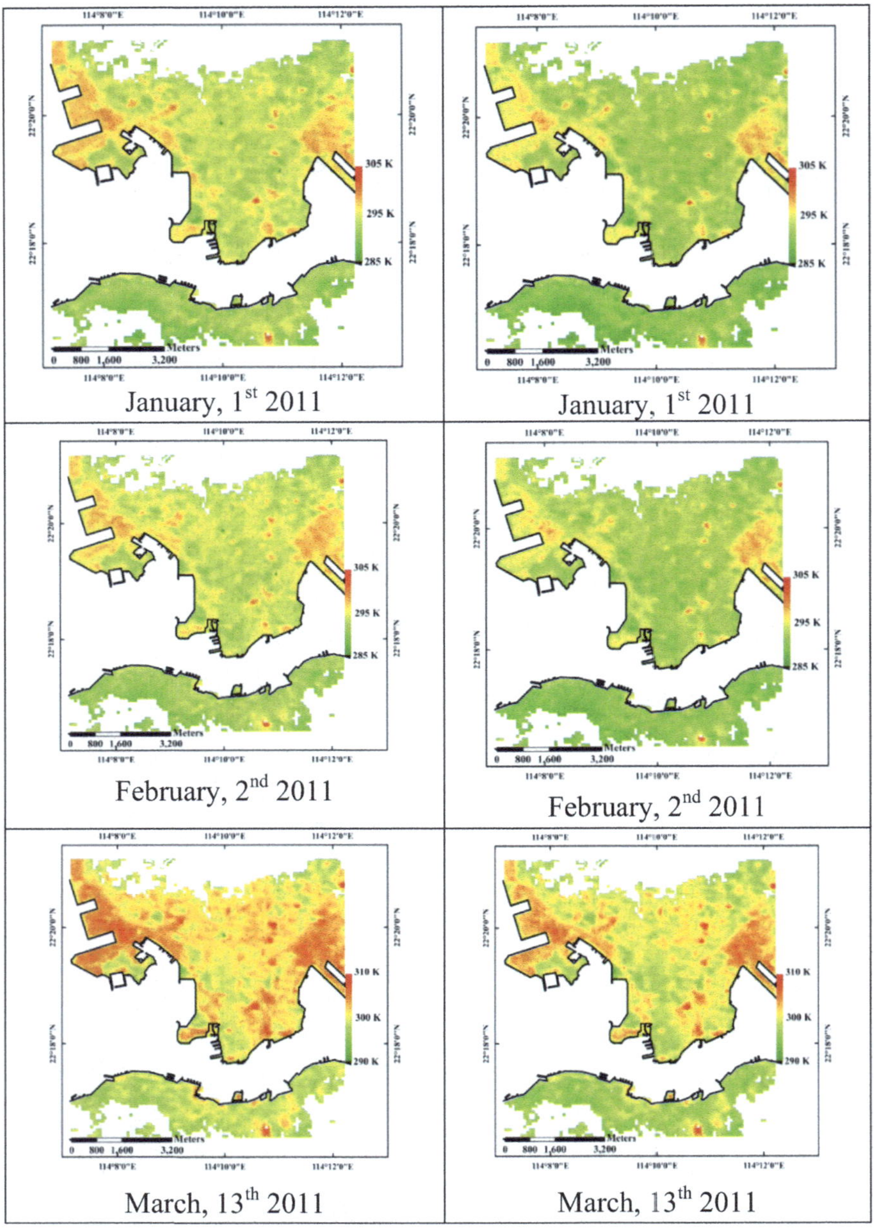

Fig. 5.10 (continued)

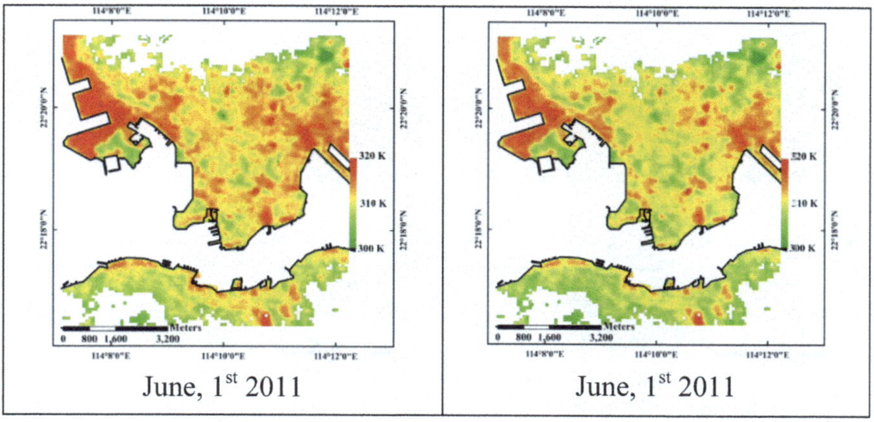

Fig. 5.10 (continued)

overestimated as surface temperatures of streets or roads are highly affected by neighboring buildings.

As thermal heterogeneity is an important factor for UST retrieval in summer its effects shall be studied by combining with the multi-angular remote sensing data in the future study.

5.5 Discussion

5.5.1 Impacts of Different Retrieval Methods on UHI

Two case studies significantly demonstrated the difference in results in retrieving emissivity and surface temperature with and without considering urban geometric effects. These results are particularly important because such differences can influence the predictions of UHI, as the accurate prediction of UHI is essential for sustainable planning and for climate change adaptation. Such differences are more complicated in a daytime scenario, in which the USTs over urban built-up areas during the daytime can be lower than those on the flat impervious surface (Fig. 5.10). High building density can block the solar irradiance resulting in generally lower temperatures over built-up areas than on the flat impervious surfaces. Such influences may also cause Urban Cool Island phenomenon [27].

Many studies used remote sensing data for SUHI research without considering geometric effects [14, 23, 28]. In order to evaluate the impacts of error of estimating on the SUHI intensity, the USTs with and without considering geometric effects from Landsat 5 in 2010 and 2011 were studied. The forest areas around the urban areas were chosen as reference. The surface temperature over urban and rural areas was first corrected to temperature of sea level based on the adiabatic lapse rate

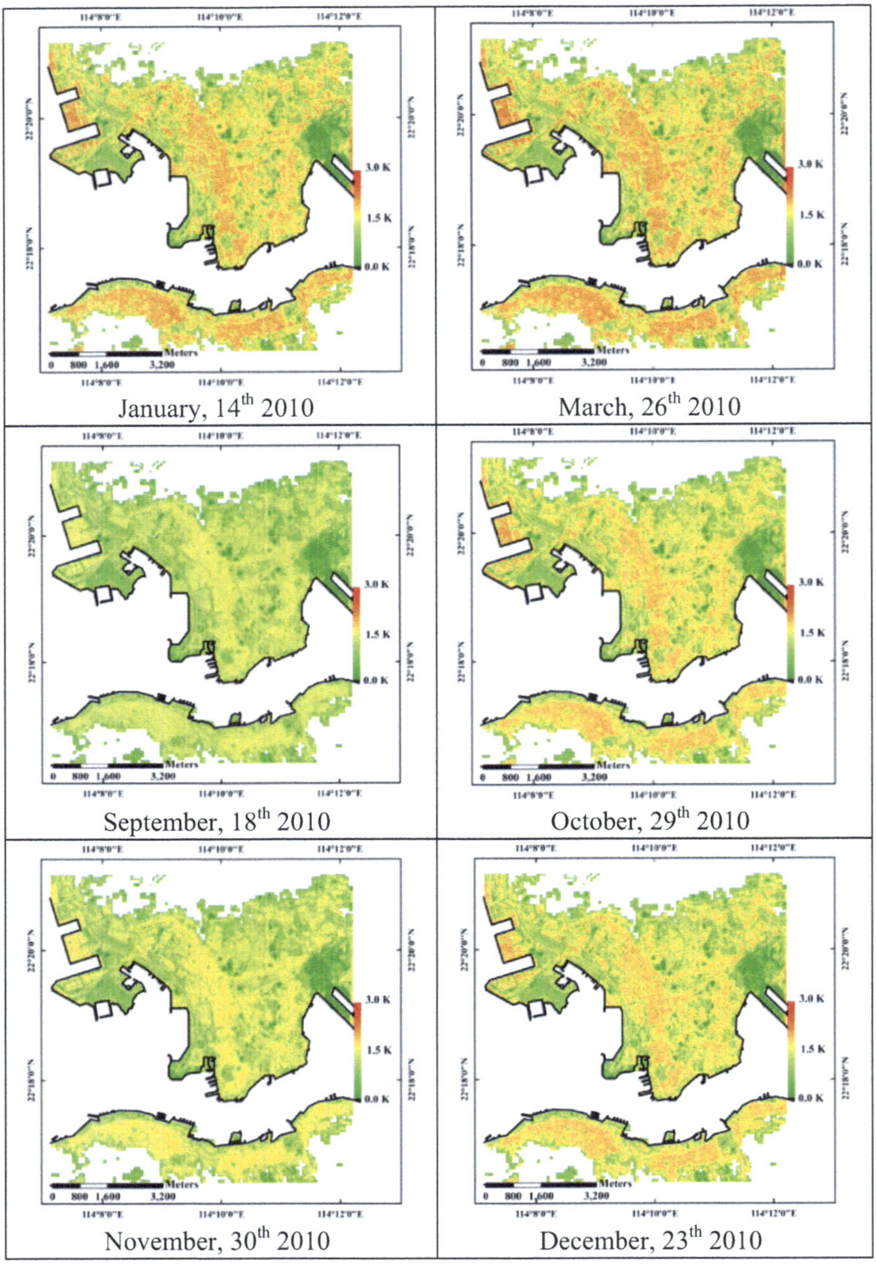

Fig. 5.11 Difference between the UST without considering geometric effects and the UST derived from UEM-SCM

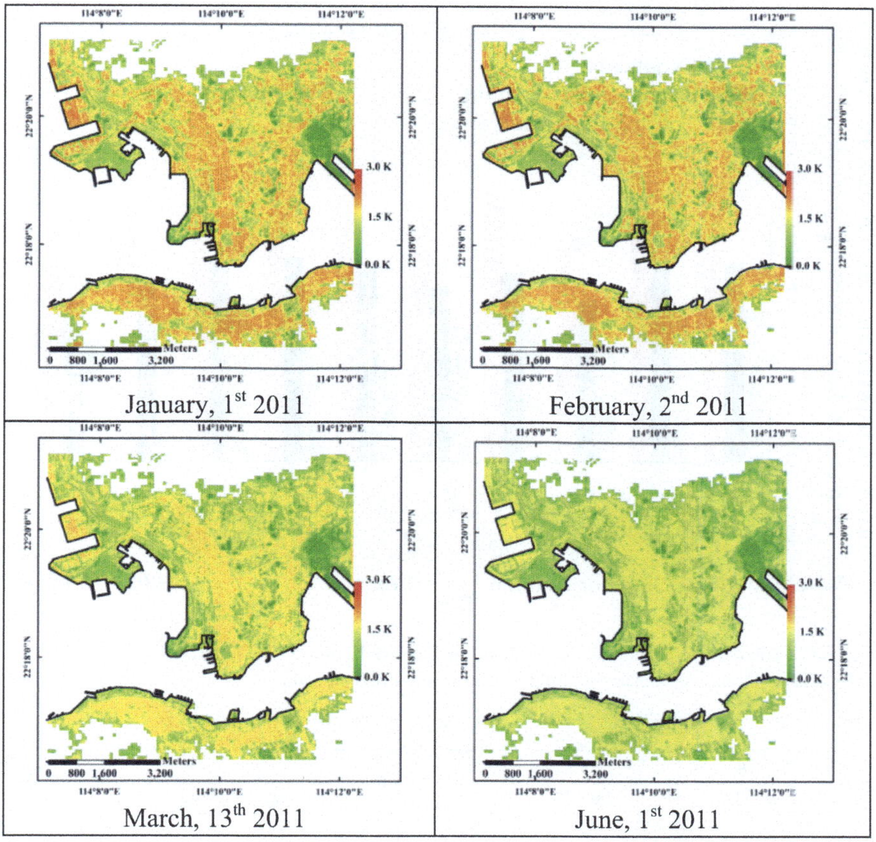

Fig. 5.11 (continued)

(−6.5 K/km), then the corrected temperatures were used to calculate the SUHI intensity. Table 5.3 shows the mean SUHI intensity based on two sets of USTs. The relative deviation of SUHI intensity based on different USTs (with and without consideration to geometric effects) can reach to 26.5% in winter (2nd February 2011) and 13.4% in summer and autumn (29th October 2010 and 1st June 2011). Since the geometric effects are more significant over built-up areas and the SUHI intensity over built-up areas at daytime is relatively lower than that over flat surfaces, the bias of SUHI intensity over built-up areas can be larger (more than 30% over built-up areas in winter) if the geometric effects are not considered in the UST retrieval.

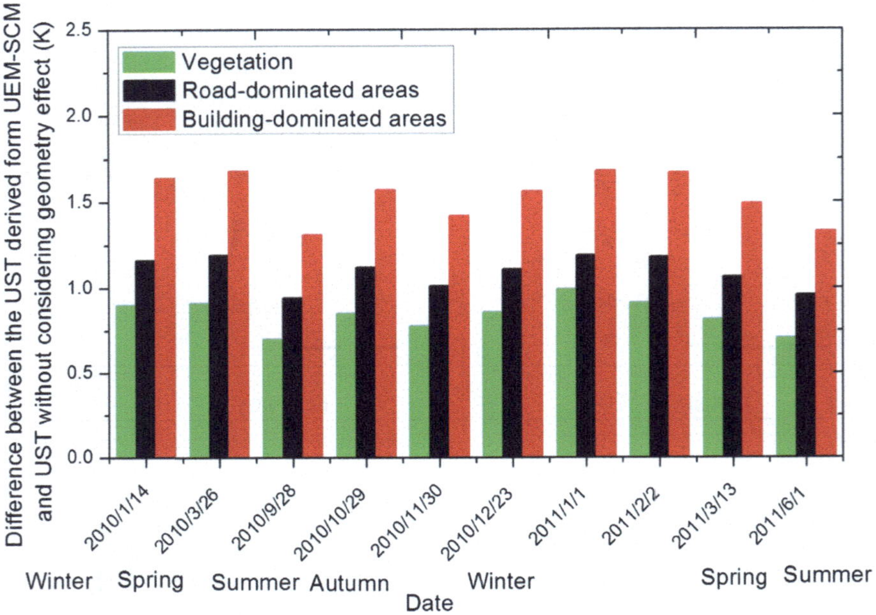

Fig. 5.12 Histogram of the average values of difference between USTs derived from UEM-SCM and USTs without considering geometric effects for Landsat 5 data

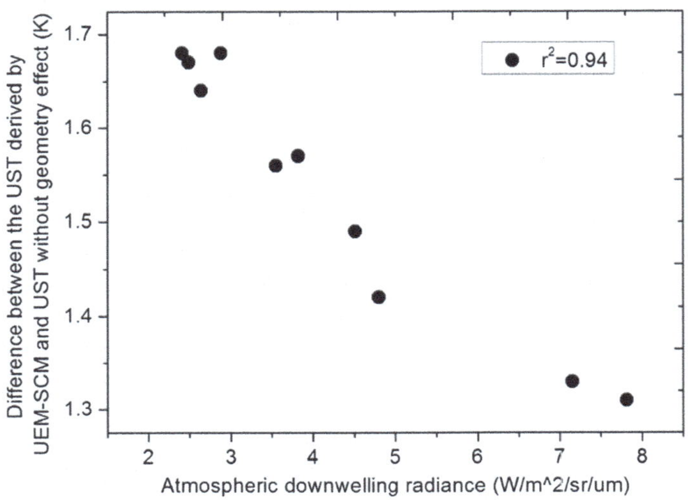

Fig. 5.13 Relation between the atmospheric downwelling radiance and the difference between two sets USTs

Table 5.3 SHUI intensity based on Landsat 5

Dates	Mean SUHI intensity (K)		Relative deviation (%)
	With considering geometric effects	Without considering geometric effects	
2010/01/14	4.66	5.89	20.9
2010/03/26	6.34	7.57	16.2
2010/10/29	7.03	8.13	13.4
2010/11/30	4.31	5.38	20.0
2010/12/23	4.35	5.53	21.3
2011/01/01	4.66	5.92	21.3
2011/02/02	3.60	4.89	26.5
2011/03/13	3.79	4.93	23.7
2011/06/01	7.11	8.25	13.8

5.5.2 Limitations

The emissivity of the ASTER products contains the urban geometric effects and effects of thermal heterogeneity within mixed pixels. However, the emissivity from the ASTER products is generally much lower than the effective emissivity derived from the IUEM-SVF. One reason is that the material emissivity used in IUEM-SVF is not perfectly accurate since it is based on the classification and building data. It is impossible to obtain accurate information on material emissivity for the whole urban areas.

The empirical relationship used in ASTER TES algorithm was established by 86 emissivity spectra of nature surfaces [6]. The actual spectral contrast over urban areas does not comply with the relationship over natural surfaces, i.e. the urban geometric effects induce significant changes on spectral shape and contrast of the urban canopy. Therefore, it is useful to build a new empirical relationship under different conditions of urban geometric characteristics and surface materials.

Moreover, the TES algorithm is based on the relationship between spectral contrast and ratio. The ASTER emissivity data product captures information of the thermal and material heterogeneity, which has complex impacts on spectral contrasts of different thermal bands, over urban areas. This makes the emissivity from the ASTER product, which is based on the TES algorithm varies over time and thus the emissivity over the same location may have different values at different times.

The UEM-SCM accounts for geometric effects using the Sky View Factor generated from DSM data. The difference between the USTs without consideration to geometric effects and the USTs with consideration to geometric effects is about 2–3 K over built-up areas.

The approach developed in this study considers geometric effects under the assumption that the mixed pixel is isothermal and does not consider the directional anisotropy of radiance observed by remote sensors caused by different observation angles and heterogeneity of component temperatures. Heterogeneous thermal distribution has an important impact on the radiative fluxes and the directional anisotropy of thermal radiation in the mixed pixel. These limit the applicability of the UEM-SCM and TES algorithm over urban areas. When the pixels are thermally heterogeneous, UEM-SCM can reduce the bias by including the geometric effects in effective emissivity. Considering that the UEM-SCM does not depend on the difference or contrast of the radiances among different thermal bands, the UEM-SCM may be more stable than TES algorithm over urban areas.

As reported by Peng et al. [18], the annual mean SUHI intensity (SUHII) in many megacities, e.g. Tokyo (Japan), Nagoya (Japan), Sao Paulo (Brazil), and Bogotá (Colombia), is approximately less than 7 K during daytime and less than 3.4 K nighttime. In Hong Kong the SUHII is about 4 K during nighttime [15]. The geometric effects can cause 2–3 K bias in the retrieved surface temperature over built-up areas. This may cause significant bias in the estimated SUHII. This indicates that the geometric effects cannot be neglected in the urban climatic research.

5.6 Conclusions

Urban surface temperature from remote sensing data can be a good option for assessing and understanding the SUHI effects, while the complex geometry limits the application of remote sensing data over urban areas. An urban effective emissivity model, IUEM-SVF, which includes the geometric effects into the emissivity estimation, was used in this study. This study first evaluated the IUEM-SVF by ASTER emissivity products at different seasons and results showed that the ε_{min}-*MMD* relationship should be reconstructed by taking urban geometric characteristics into account to improve the accuracy of urban surface temperature retrieval. The urban surface temperatures with and without consideration to geometric effects were retrieved based on single channel method. Results showed that geometric effects over urban areas varied with seasons, where urban surface temperature retrieval in winter were larger than that in summer. Finally, the different urban surface temperatures, i.e. with and without considering geometric effects were used to estimate SUHI intensity, and results showed that different urban surface temperatures can cause 26.5% bias in winter and 13.4% bias in summer and autumn for estimating SUHI intensity.

Acknowledgements This work was supported in part by the grant of Early Career Scheme (project id: 25201614) and General Research Fund (project id: 515513) from the Research Grants Council of Hong Kong; the grant 1-ZE24 from the Hong Kong Polytechnic University. The authors thank the Hong Kong Planning Department, the Hong Kong Lands Department, the Hong Kong Civil Engineering and Development Department, and the Hong Kong Observatory for the planning, building GIS, weather and climate, and airborne LiDAR data, and NASA LP DAAC for the ASTER and Landsat satellite imagery.

References

1. Amiri R, Weng Q, Alimohammadi A, Alavipanah SK (2009) Spatial–temporal dynamics of land surface temperature in relation to fractional vegetation cover and land use/cover in the Tabriz urban area, Iran. Remote Sens Environ 113(12):2606–2617
2. Chakraborty SD, Kant Y, Bharath BD (2014) Study of land surface temperature in delhi city to managing the thermal effect on urban developments. Int J Adv Sci Tech Res 4(1):439–450
3. Chen L, Ng E, An X, Ren C, Lee M, Wang U, He Z (2012) Sky view factor analysis of street canyons and its implications for daytime intra-urban air temperature differentials in high-rise, high-density urban areas of Hong Kong: a GIS-based simulation approach. Int J Climatol 32 (1):121–136
4. Cheng KS, Su YF, Kuo FT, Hung WC, Chiang JL (2008) Assessing the effect of landcover changes on air temperature using remote sensing images—a pilot study in northern Taiwan. Landsc Urban Plan 85(2):85–96
5. Dousset B, Gourmelon F (2003) Satellite multi-sensor data analysis of urban surface temperatures and landcover. ISPRS J Photogramm Remote Sens 58(1–2):43–54
6. Gillespie A, Rokugawa S, Matsunaga T, Cothern JS, Hook S, Kahle AB (1998) A temperature and emissivity separation algorithm for advanced spaceborne thermal emission and reflection radiometer (ASTER) images. IEEE Trans Geosci Remote Sens 36(4): 1113–1126
7. Hu L, Brunsell NA (2013) The impact of temporal aggregation of land surface temperature data for surface urban heat island (SUHI) monitoring. Remote Sens Environ 134:162–174
8. Kotthaus S, Smith TEL, Wooster MJ, Grimmond CSB (2014) Derivation of an urban materials spectral library through emittance and reflectance spectroscopy. ISPRS J Photogramm Remote Sens 94:194–212
9. Krayenhoff ES, Voogt J (2007) A microscale three-dimensional urban energy balance model for studying surface temperatures. Bound-Layer Meteorol 123(3):433–461
10. Lagouarde J-P, Hénon A, Irvine M, Voogt J, Pigeon G, Moreau P, Masson V, Mestayer P (2012) Experimental characterization and modelling of the nighttime directional anisotropy of thermal infrared measurements over an urban area: case study of Toulouse (France). Remote Sens Environ 117:19–33
11. Lagouarde JP, Irvine M (2008) Directional anisotropy in thermal infrared measurements over Toulouse city centre during the CAPITOUL measurement campaigns: first results. Meteorol Atmos Phys 102(3–4):173–185
12. Lai A, So AC, Ng S, Jonas D (2012) The territory-wide airborne light detection and ranging survey for the Hong Kong special administrative region, In: The 33RD Asian conference on remote sensing, pp 26–30
13. Li Z-L, Tang B-H, Wu H, Ren H, Yan G, Wan Z, Trigo IF, Sobrino JA (2013) Satellite-derived land surface temperature: current status and perspectives. Remote Sens Environ 131:14–37
14. Liu L, Zhang Y (2011) Urban heat island analysis using the Landsat TM data and ASTER data: a case study in Hong Kong. Remote Sens 3(7):1535–1552
15. Nichol JE, Fung WY, Lam K-S, Wong MS (2009) Urban heat island diagnosis using ASTER satellite images and 'in situ' air temperature. Atmos Res 94(2):276–284
16. Oltra CR, Cubero-Castan M, Briottet X, Sobrino JA (2014) Analysis of the performance of the TES algorithm over urban areas. IEEE Trans Geosci Remote Sens 52(11):6989–6998
17. Payan V, Royer A (2004) Analysis of temperature emissivity separation (TES) algorithm applicability and sensitivity. Int J Remote Sens 25(1):15–37
18. Peng S, Piao S, Ciais P, Friedlingstein P, Ottle C, Bréon F-M, Nan H, Zhou L, Myneni RB (2012) Surface urban heat Island across 419 global big cities. Environ Sci Technol 46(2): 696–703
19. Revi A, Satterthwaite DE, Aragón-Durand F, Corfee-Morlot J, Kiunsi RB, Pelling M, Roberts DC, Solecki W (2014) Urban areas. Climate Change 535–612

20. Soux A, Voogt JA, Oke T (2004) A model to calculate what a remote sensorsees' of an urban surface. Bound-Layer Meteorol 111(1):109–132
21. Voogt JA (2008) Assessment of an urban sensor view model for thermal anisotropy. Remote Sens Environ 112(2):482–495
22. Voogt JA, Oke TR (2003) Thermal remote sensing of urban climates. Remote Sens Environ 86(3):370–384
23. Weng Q (2003) Fractal analysis of satellite-detected urban heat island effect. Photogramm Eng Remote Sens 69(5):555–566
24. Yang J, Wong MS, Menenti M, Nichol J (2015) Modeling the effective emissivity of the urban canopy using sky view factor. ISPRS J Photogramm Remote Sens 105:211–219
25. Yang J, Wong MS, Menenti M, Nichol J (2015) Study of the geometry effect on land surface temperature retrieval in urban environment. ISPRS J Photogramm Remote Sens 109:77–87
26. Yang J, Wong MS, Menenti M, Nichol J, Voogt J, Krayenhoff ES, Chan PW (2016) Development of an improved urban emissivity model based on sky view factor for retrieving effective emissivity and surface temperature over urban areas. ISPRS J Photogramm Remote Sens 122:30–40
27. Yang X, Li Y, Luo Z, Chan PW (2017) The urban cool island phenomenon in a high-rise high-density city and its mechanisms. Int J Climatol 37(2):890–904
28. Yuan F, Bauer ME (2007) Comparison of impervious surface area and normalized difference vegetation index as indicators of surface urban heat island effects in Landsat imagery. Remote Sens Environ 106(3):375–386

The manufacturer's authorised representative in the EU is Springer
Nature Customer Service Centre GmbH, Europaplatz 3, 69115 Heidelberg,
Germany. If you have any concerns regarding our products, please
contact ProductSafety@springernature.com

Printed and bound by CPI Group (UK) Ltd, Croydon, CR0 4YY
25/09/2026
02222891-0017